Sidney Barwise

The Purification of Sewage

Being a Brief Account of the Scientific Principles of Sewage Purification and their

Practical Application

Sidney Barwise

The Purification of Sewage
Being a Brief Account of the Scientific Principles of Sewage Purification and their Practical Application

ISBN/EAN: 9783337418298

Printed in Europe, USA, Canada, Australia, Japan

Cover: Foto ©berggeist007 / pixelio.de

More available books at **www.hansebooks.com**

THE PURIFICATION OF SEWAGE

BEING

*A BRIEF ACCOUNT OF
THE SCIENTIFIC PRINCIPLES OF SEWAGE PURIFICATION
AND THEIR PRACTICAL APPLICATION*

BY

SIDNEY BARWISE, M.D.(Lond.)

M.R.C.S. D.P.H.(Camb.)

FELLOW OF THE SANITARY INSTITUTE
MEDICAL OFFICER OF HEALTH TO THE DERBYSHIRE COUNTY COUNCIL

LONDON
CROSBY LOCKWOOD AND SON
7, STATIONERS' HALL COURT, LUDGATE HILL
1899

PREFACE.

In the present volume—a small book upon a large subject—the question of the Purification of Sewage is dealt with chiefly from a chemical and biological point of view, and in the light of the experience gained in the discharge of my duties as the Medical Officer of Health of a large County. That office has afforded me constant opportunities for the inspection of sewage works in actual operation, and for analyzing the effluents from such works. As the result, I have endeavoured to set out in these pages as succinctly as possible, and yet (it is hoped) with sufficient fulness for the purpose in view, the conditions which appear favourable for particular processes for the purification of sewage, and their necessary limitations.

Until the passing of the Local Government Act, 1888, by which County Councils were constituted, the enforcement of the Rivers Pollution Prevention Act of 1876 was left in the hands of the Sanitary Authorities, who, being themselves the chief offenders against the provisions of that enactment, were, naturally enough, very loth to institute proceedings against one another.

Now, however, under pressure from County Councils and Joint River Boards, the District Councils are seriously taking in hand the work of saving our rivers and streams from pollution, and the question is constantly being asked, "What is the best process of sewage purification to adopt?" The form of this question shows only too clearly that the real nature of the problem is not generally understood. It is, indeed, of the first importance that members and officers of the various bodies, upon whom devolves the duty of dealing with the problem of purification of sewage, should grasp the scientific principles of the question, and recognize the fact that the particular scheme to be adopted in any locality must depend upon a variety of local circumstances. It is with the view of giving some help in this direction that the following pages have been written.

During recent years, by means of numerous bacteriological and chemical investigations, both in this country and in America, great advances have been made in our knowledge of the changes which sewage undergoes in purification, and not a few conclusions of wide-reaching importance established; and it is hoped that the presentation in this little work of some of the results thus attained will be found useful by engineers and other officials who wish to avail themselves of the latest researches of chemists and biologists upon the question of Sewage Purification.

SIDNEY BARWISE.

COUNTY OFFICES, DERBY,
October, 1898.

CONTENTS.

CHAPTER I.

SEWAGE: ITS NATURE AND COMPOSITION.

PAGE

Definition of sewage—Sewage from water-closet and non-water-closet towns—Plea for the general adoption of water-closets   1

CHAPTER II.

THE CHEMISTRY OF SEWAGE.

Variations of sewage with water supply—Composition of average sewage—Explanation of chemical terms used—Total solids, free ammonia, organic or albuminoid ammonia, chlorine, oxygen absorbed—Formula for calculating flow—Standards of purity for sewage effluents—Chemical changes to be effected by purification   10

CHAPTER III.

VARIETIES OF SEWAGE AND THE CHANGES IT UNDERGOES.

Chemical composition of different sewages—Variations in the chlorine—Night and day sewage—Hourly variations in flow and composition—Changes sewage undergoes by keeping—Purification effected by clarification, nitrification, and bacteriolysis—Importance of sewers being self-cleansing—Table giving maximum discharge of various sized sewers at different inclinations 28

CHAPTER IV.

RIVER POLLUTION AND ITS EFFECTS.

Self-purification of rivers—Bacterial purification, by sedimentation, by biological changes—Gross pollution causing nuisance—Ptomain poisoning from polluted waters—Effects of polluted water on oyster culture—Typhoid fever and polluted river water 41

CHAPTER V.

THE LAND TREATMENT OF SEWAGE.

Sewage farms—Only indicated where soil is suitable—Clay lands nearly useless—The theory of nitrification—Intermittent downward filtration—Irrigation proper—Stratford-on-Avon farm—Edinburgh farm—Paris farm—Berlin farm—Proper method of laying out and working a farm—Quantity of land required—Fallacy of the manurial value of sewage ... 53

CHAPTER VI.

PRECIPITATION, PRECIPITANTS, AND TANKS.

Limitations of the process—In practice solid matters only removed—Soluble matter removed by excess of precipitants—Comparative costs of precipitants—Lime, alum, coperas

CONTENTS. xi

PAGE

—Various patent precipitants—Precipitation without tanks —Absolute rest tanks—Continuous flow tanks—Dortmund tank of Kinebühler—The Candy tank—Sludge and methods of dealing with it 85

CHAPTER VII.

FILTRATION OR NITRIFICATION.

Intermittent land filtration—Artificial filters suggested by Warington in 1882—Conditions necessary for nitrification—Experiments by the Massachusetts Board of Health with coarse sand, fine sand, peat, silt, garden soil, and fine gravel—Results obtained—London County Council experiments—Results obtained 106

CHAPTER VIII.

SPECIAL FORMS OF SEWAGE FILTERS.

Ducat's filter—Garfield's coal filter—Sewage distributors—The Lowcock filter—Comparative results with coke, breeze, coal, sand, and Lowcock's filter, Tipton and Buxton—Automatic arrangements for applying sewage intermittently ... 123

CHAPTER IX.

THE NEW DEPARTURE: "BACTERIOLYSIS."

Scott Moncrieff's and Adeney's experiments—The Exeter septic process of Cameron—Results obtained at Exeter—Gases resulting from the septic process 134

CHAPTER X.

CONCLUSION.

What process of purification to adopt—When a sewage farm—When precipitation and land—When precipitation and artificial filtration—Indications for a septic process 141

INDEX 147

THE PURIFICATION OF SEWAGE.

CHAPTER I.

SEWAGE: ITS NATURE AND COMPOSITION.

Definition of sewage—Sewage from water-closet and non-water-closet towns—Plea for the general adoption of water-closets.

By "sewage," which is derived from the Anglo-Saxon "seon," to flow down, is meant the liquid contents of a sewer. This, in its simplest form from a village or non-water-closet town, is frequently designated "slop-water." It contains the liquid excretions of the inhabitants; the foul waters from the kitchens containing vegetable and animal matters, bits of fat, and other refuse; the "suds" from the washing of dirty linen, cooking utensils, and the people themselves, holding in solution and suspension soap, fatty acids, and the exudations from the human skin. Such soapy slops, as every one is aware, if allowed to stand for twenty-four hours, become most foul and offensive. Then there is the dirty water from the washing of floors, the swilling of yards, the solid and liquid excretions of animals in the streets, the drainage from stables and pig-sties, the blood and other animal matters from slaughter-houses, silt from street sweepings, and sometimes, if the town is an old one, the most offensive and concentrated filth of all, the soakage from a privy-midden. From a consideration of its various constituents, I think it must be admitted that even what is called "slop-water" is a polluting noxious liquid, no less offensive than ordinary sewage.

In the case of water-closet towns, in addition to the above polluting matters, there are the solid excreta from the inhabitants, paper and other matters of a like nature emptied through the closets into the sewers, but there is also a larger amount of clean water. As a rule, in both cases, the surface water from the streets and from the yards, and a certain amount of ground water, finds its way into the sewers.

Sewage as defined is the liquid contents of a sewer, and as a sewer has a specific meaning under the Public Health Acts, the definition of this word becomes a point of great importance, involving, as it does, an answer to the question, What is sewage?

By sect. 4 of the Public Health Act, 1875, a "drain" is defined as meaning any drain of and used for the "drainage of one building only," or premises within the same curtilage, and made merely for the purpose of communicating therefrom with a cesspool or with a sewer; whilst "sewer" includes sewers and drains of every description, except those to which the word "drain" interpreted as above applies. From this definition it follows that where two or more houses have a common drain, that drain is, within the meaning of the Public Health Act, 1875, a sewer, and as such, by sect. 13 of the same Act, it is vested in the Sanitary Authority, who become, by sect. 3 of the Rivers Pollution Prevention Act, 1876, responsible for using "the best practicable and available means to render harmless the sewage matter" falling or flowing therefrom into any stream or watercourse.

The importance of this definition can hardly be exaggerated, and numerous cases have arisen on the point. The decisions of the Court of Appeal, however, have made it quite clear that where two or more houses drain into a common channel, whether that channel be a pipe-drain, or even an open ditch,* it is a sewer, and the Sanitary Authority are liable for its maintenance in such a state as not to be a nuisance, to keep it properly cleansed, ventilated, and repaired (sects. 15 and 19 Public Health Act, 1875), and to use the best practicable and available means to purify the sewage flowing from it.

Sewage is an extremely complex fluid, varying in the same place from hour to hour both in its volume and in its chemical composition, nor is the sewage

* *Wheatcroft* v. *Matlock Local Board*, 52 L. T. (N.S.) 356.

of any two places the same, even though it be in each case of a purely domestic character.

Dealing first of all with the sewage from non-water-closet towns: the greatest cause of its variation is the quantity of water per head supplied to the district. It is frequently supposed that this sewage is not of such a polluting nature as that from towns where the water-carriage system has been adopted. Although the total amount of sewage matter conveyed to the outfall by the sewers is greater where water-closets are in use, yet the average composition of such sewage is no stronger than that from towns where there are no water-closets. The reason for this is that water-closets are not generally used except in towns where there is an ample water supply; and although it is true that by the general adoption of water-closets two and a half ounces of excrement per head per day, containing 25 per cent. of solid matter, are added to the sewage, yet the water which is used for flushing, and the unavoidable waste which results from the adoption of the water-carriage system, are sufficient to dilute this five-eighths of an ounce of solid matter below the strength of average sewage from a non-water-closet town.

With regard to the sewage of water-closet towns, the chief difference is that it contains a larger amount of suspended matter, such as disintegrated paper, the filaments of which are liable, if not removed, to form a film of papier-maché on the surface of sewage filters. In the manufacturing towns there is also more or less refuse from factories draining into the sewers.

By sect. 7 of the Rivers Pollution Prevention Act, 1876, every sanitary authority is compelled to give facilities for enabling manufacturers within its district to carry the liquids proceeding from their factories into the public sewers, provided that such liquids do not prejudicially affect the sewers or the system of disposal of the sewage, or are not from their temperature or otherwise injurious from a sanitary point of view. The chief manufacturing processes which produce liquid wastes, that manufacturers may ask facilities for draining into the sewers, are brewing, paper making, bleaching, calico printing and dyeing, wool scouring, fulling and felting, tanning and fell-mongering, and chemical and alkali works. Some of these wastes if admitted in a reasonable proportion are advantageous, while other, such as galvanizers' pickle, may, unless special precautions are taken, be injurious to the sewers, and also to the system of sewage disposal.

Where manufacturing wastes are admitted into the sewers, arrangements should be made for their admission evenly and continuously. It is no uncommon thing to see the sewage of a manufacturing town completely change in its appearance and composition several times in the course of an hour, defeating all attempts at purification.

Water-closets.—The sewage from water-closet towns is no stronger than ordinary sewage or slop-water, which is a polluting liquid necessitating purification works whether the excreta of the inhabitants are added to the sewage or not. This being so, the advantage of adopting water-closets is apparent. The alternatives are the pail system or some form of privy-midden.

With regard to the pail system, the cost of emptying a pail weekly is on the average about eight shillings per annum per house, or twopence a week. On the other hand, if water can be obtained at eightpence a thousand gallons, the cost in water for each house is about four shillings per annum.

As for the midden system, it is now too late, at the beginning of the twentieth century, to contemplate a system which intensifies all the evils of the pail system, and is without any of its advantages. In rural districts, where there is no public water supply, it may still linger; but the difficulty of getting privies regularly emptied increases every year. The introduction of preserved articles of diet has led to empty tins, broken bottles, etc., being thrown into the ash-pits, and I have heard farmers declare that it has cost them as much to sort the contents of privy-middens, or to pick up the old tin cans, from their fields, as they have been paid for removing the refuse.

Where sewers have been provided, why not make

full use of them? The only system fitted for this country is that of water-closets. The water comes silently into the houses, and does its work efficiently; the excreta are removed at once before the mass has time to give rise to any nuisance by decomposing. The danger of sewer gas entering a house is just as great from a bath or sink as from a water-closet if the house drains are not disconnected from the public sewer, while there is no more difficulty in disconnecting a water-closet than there is in cutting off a bath waste. Both should be trapped, and then made to pass *through the open air*, in a disconnecting trap (see Fig. 1), before passing into the sewer. If this simple precaution is taken, and the drains are not defective, it is impossible for sewer gas to enter the house; while the risk of the drains being defective is not increased by the fact that there is a water-closet within the house.

Although this matter does not arise on the question of sewage disposal, it is of such importance that I feel I am justified in introducing a sketch showing the proper method of disconnecting water-closets, bath and sink wastes, from the sewers. This is done in Fig. 1.

That danger of typhoid fever from the adoption of the water-carriage system, if properly carried out, does not exist, is clearly demonstrated in the case of London. Here is the largest population ever congregated together, water-closets generally adopted,

8   SEWAGE PURIFICATION.

and a low typhoid rate. That water-carriage is the healthiest system has been conclusively shown by Dr. Boobyer in the case of Nottingham, the following

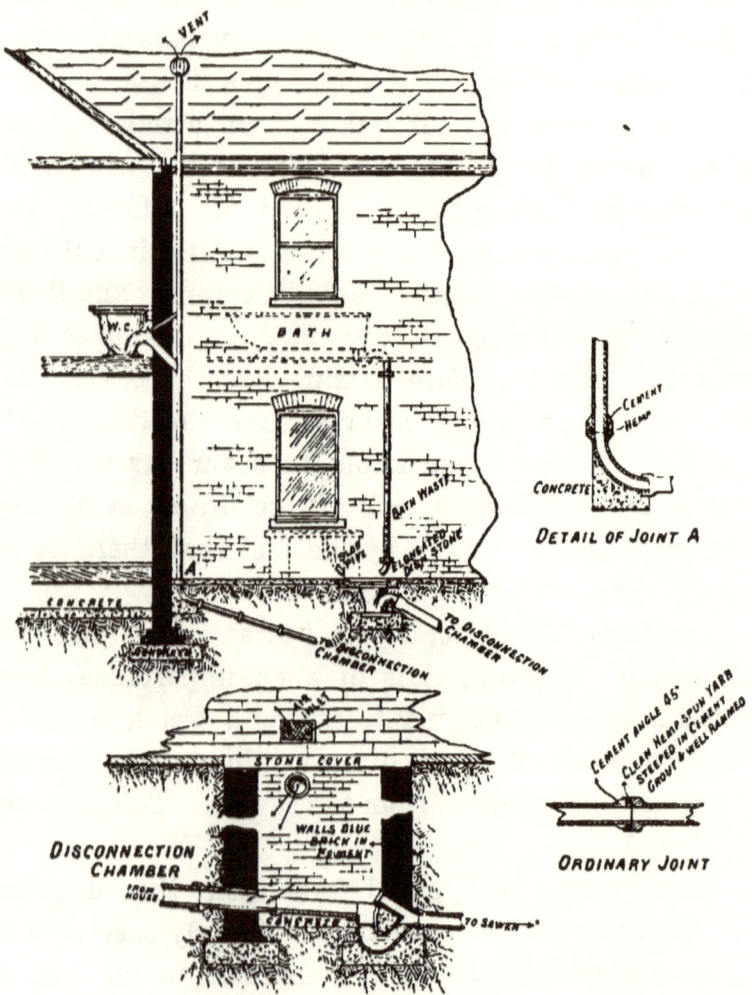

Fig. 1.—Method of disconnecting House Drains from Public Sewers.

summary giving the incidence during the last ten years of typhoid fever on houses with water-closets, pails, and privy-middens :—

    Privy-midden houses, 1 case of typhoid in 37
    Pail-closet   ,,   1   ,,   ,,   120
    Water-closet ,,   1   ,,   ,,   558

In the face of these figures, the advantage of the water-carriage system is apparent.

CHAPTER II.

THE CHEMISTRY OF SEWAGE.

Variations of sewage with water supply—Composition of average sewage—Explanation of chemical terms used—Total solids, free ammonia, organic or albuminoid ammonia, chlorine, oxygen absorbed—Formula for calculating flow—Standards of purity for sewage effluents—Chemical changes to be effected by purification.

An average sewage contains about 100 grains of solid matter to the gallon, the remaining 69,900 grains being water. Of the 100 grains of solid matter, a certain proportion, varying in different towns and at different times of the day and seasons of the year, will consist of organic matter; perhaps on the average it will amount to 40 grains. It is the object of sewage purification to remove, as far as possible, this 40 grains of organic matter, and to as nearly as possible completely oxidize that which cannot be removed. All the organic and mineral matter in suspension should be removed, and in a good effluent as much as 80 per cent. of the organic matter in solution will also be oxidized to nitrates. An average sewage, containing 100

COMPOSITION OF SEWAGE.

grains of solid matter to the gallon, will have about 70 grains in solution; this includes the solid matter in solution in the drinking water of the district. Consideration of this fact will at once show that the solid matter in solution in a sewage must vary in different towns with the quality of the drinking water. A glance at the following Table, giving the solid matter in water supplies from various sources, will show how the percentage of solid matter in a sewage varies with the source of the public water supply :—

GRAINS OF SOLID MATTER TO THE GALLON IN VARIOUS WATERS.

Formation.	Surface waters.	Deep well and spring waters.
Igneous rocks, Metamorphic, Silurian, and Devonian	3 to 7	5 to 15
Yoredale and Millstone Grits	5 to 10	15 to 28
New Red Sandstone	8 to 18	20 to 30
Mountain Limestone	8 to 16	22·5
Magnesian Limestone	8 to 18	35·0
Chalk	—	25·0
Chalk beneath London Clay	—	54·5
Coal Measures	16	60·0

It will be seen that the percentage of solid matter in solution in the sewage from a town supplied with water from a deep well in the coal measures, for instance, will be ten times greater than that from a town which is supplied with surface water from a gritstone gathering ground. It is obvious that, as far as the purification of sewage is concerned, the amount of this solid mineral matter in solution is of secondary importance. An average sample of sewage, containing 100 grains of solid matter to the gallon, will have about 70 grains of solid matter in solution and 30 in suspension. Of the suspended matter a varying amount, somewhere about two-thirds, will be organic matter, while of the other in solution less than one-third will be organic matter.

The following Table may be useful as giving an idea of the solid matter in an average sewage :—

COMPOSITION OF ONE GALLON OF TYPICAL AVERAGE SEWAGE.

	Grains per gallon.
(1) Solid matters in suspension—	
(*a*) Organic	20
(*b*) Mineral	10
	30
(2) Solid matters in solution—	
(*a*) Organic	20
(*b*) Mineral	50
	70
Total	100

A gallon of water weighs 70,000 grains, so that 70 grains per gallon is equivalent to 100 parts in the 100,000. As the decimal notation is used in chemical investigations, it is easier to express results in parts per 100,000 than in grains per gallon. In this work results will be expressed in parts per

100,000. It is a simple matter to convert grains per gallon to parts per 100,000 by dividing by 7 and multiplying by 10; similarly parts per 100,000 can be converted to grains per gallon by multiplying by 7 and dividing by 10.

Such a sewage as the above would, upon analysis, probably yield results somewhat as follows, expressed in round numbers in parts per 100,000 :—

Total solids.	Solids in suspension.	Chlorine.	Free ammonia.	Organic ammonia.
142·8	42·8	12·0	5·0	1·0

Nearly all the solid matter in suspension and a fraction of that in solution can be removed by almost any precipitating process, such processes about half purifying the sewage. In the laboratory the solid matter in suspension is removed by passing the sewage through filter paper. The quantity can be thus estimated. After the removal of the suspended matter, there is still the 20 grains of organic matter in solution to purify. As we shall see, this organic matter in solution, having as its chief source the urine which enters the sewers, contains the bulk of the nitrogen in the sewage and cannot be removed. Its purification can only be effected by completely oxidizing it, so that when it enters a river its affinity for oxygen is satisfied and it does not deprive the river of its normal oxygen which is required for the natural fauna and flora of the river water.

The chemical terms which are used in relation to

sewage are various, but the following are the only terms we need trouble with, as they indicate the only determinations which it is necessary should be made.

Total Solids.—The meaning of the term "solid matter" has been fully explained. It is determined by evaporating to dryness in a platinum dish a known quantity of sewage over a water bath, drying at 212° Fahr., and weighing immediately on cooling. The result obtained is the total solid matter. The total organic matter is then determined by igniting the residue in the platinum dish and weighing again; the loss on the first weighing gives the total amount of organic matter which has been burnt off. The sewage is then filtered, and the same process is gone through again with the filtered sewage; the results obtained will give the total solids and the organic solids in solution. By subtracting these from the first results the total suspended matter and the organic suspended matter will be obtained. A good sewage effluent should contain no organic matter in suspension, and not more than three parts per 100,000 of organic matter in solution.

Free Ammonia.—This is the ammonia formed from the splitting up of the nitrogenous matters in the sewage by the action of bacteria. There are at least two varieties of bacteria which split up the urea in the urine and convert it into carbonate of ammonia. Although the term "free ammonia" is used, the ammonia is not to be understood to exist in the free state, using the word "free" in its chemical sense. It is in combination with carbonic and organic acids, and if mineral acids enter the sewers it may also exist as the chloride or the sulphate of ammonium. The term "bacteriolysis" has been proposed for this splitting up of organic matter by the action of living organisms. The free ammonia in itself is quite harmless and is not of much importance.

The Organic or Albuminoid Ammonia and the Organic Nitrogen.—The organic matter of animal origin both in solution and in suspension contains about sixteen per cent. of nitrogen; the vegetable organic matter contains much less. At one time the organic nitrogen was estimated as an index of the amount of organic matter. The process, however, has not been generally adopted.

Wanklyn showed that by boiling organic matter with permanganate of potash and caustic potash it was split up, and a proportion of the nitrogen was given off in the form of ammonia, which could easily be estimated, and was as reliable as an index of the organic matter as the organic nitrogen proposed by the old Rivers Pollution Commissioners. The terms "albuminoid" and "organic ammonia" are synonymous. As a rule, about two-sevenths of the organic nitrogen in sewage will come off as organic ammonia. This is perhaps the most important determination that can be made of sewage and sewage effluents. The organic ammonia in a sewage varies from 3 or even more parts per 100,000, to as little as 0·2 part per 100,000, while that in a sewage effluent should not yield more than 0·1 part per 100,000. The Table on page 29 gives the results of the average of a number of samples of sewage from different sources. It will be seen that the American sewage is very much weaker than average English sewage.

Chlorine.—The chlorine in sewage has its origin in common salt, of which it constitutes about 60 per cent. It enters the sewage as salt in the waste from kitchens and in the urine. There is also a certain amount of chlorine in all drinking water. It is also present, to a slight extent, in rain water. On the West Coast as much as 5 parts per 100,000 have been found in rain water; at Liverpool 1 part; and in various parts of England (inland) about 0·3 per 100,000. The chlorine in drinking waters varies with the source of the water; in upland surface waters it is about 1 part in 100,000, while in the deep-well waters from the millstone grit, new red sandstone, mountain limestone, chalk, Devonian rocks, it is about 2 or 3 parts; while in the chalk beneath the London clay it is as much as 15, and in the coal measures even 18 parts per 100,000. So that the amount of chlorine in sewage will very materially depend upon the drinking water supplied to the district.

Human urine, of which on the average, in a mixed

population, about 40 ounces per head per day are excreted, contains about 842 parts per 100,000 of common salt, of which 500 parts consist of chlorine. The chlorine itself is of little importance; there is no necessity for its removal, and *no process of purification can remove it*. Its determination, however, is important, because it serves as a valuable index of the strength of ordinary domestic sewage. For instance, one frequently sees in the advertisements of patented processes of purification, an analysis published of the wonderful purity of a sewage effluent which contains a small amount of chlorine, and beside it an analysis of the raw sewage which contains a much larger amount of chlorine. Obviously the effluent and sewage do not correspond, and the effluent has been obtained from a weaker raw sewage than that of which the analysis is published.*

From this it will be obvious that the determination of chlorine is always an important one, as, if the percentage of purification effected by any process is to be gauged, the chlorine in the effluent must be practically the same as that in the sewage. It is a

* In a paper read at the Sanitary Institute Congress, Liverpool, in 1894, on a "patented" process of purification, the chlorine in a raw sewage was given as 30 parts per 100,000; while that in the effluent from the "patent" filters was given as 3·9 parts per 100,000. The effluent must have been from a sewage of between one-seventh and one-eighth of the strength of that purporting to be the sewage before purification.

pity that this elementary fact has not been more generally appreciated by the advocates of many patented processes, as the analyses they publish prove too much.

The chlorine in twenty samples of sewage from midden towns was found by the Rivers Pollution Commissioners to contain 11·54 parts per 100,000; while that from 36 water-closet towns contained 10·66 parts. The drainage from districts where the wash water after dyeing enters the public sewers contains less chlorine than average sewage; while the drainage from 15 districts where bleaching, fulling, and scouring were carried on was shown by the Rivers Pollution Commissioners to contain 20 parts per 100,000; the drainage from paper mills contained about the same. The drainage from alkali works contained from 100 to 700 parts; while that from galvanizing and tin-plate works on the average contains 2000 parts, and may contain as much as 20,000 parts per 100,000. From this it will be seen that the amount of chlorine in the sewage will depend upon the nature of the manufacturing refuse.

If the sewage is purely of a domestic nature, and there is a public water supply, the amount of chlorine in it being known, the strength of the sewage and the number of gallons per head can be approximately ascertained by estimating the quantity of chlorine in the sewage. The following is a formula

CHLORINE.

which the author has devised for this purpose, and which gives approximately accurate results:—

Let A = parts per 100,000 of chlorine in the public water supply, and let B = the parts per 100,000 of chlorine in the sewage; let X be the number of gallons of sewage per head per day required to be ascertained—

$$X = \frac{125}{B - A}$$

For example, on a recent occasion the chlorine in the Buxton drinking water was found to be 1·1 parts per 100,000 (A) = 1·1, and the chlorine in the sewage to be 2·3 (B) = 2·3:

$$X = \frac{125}{2·3 - 1·1}$$

= 104 gallons per head per day. As a matter of fact, the flow was actually about 100 gallons per head per day.

The Oxygen absorbed in Four Hours at Eighty Degrees Fahrenheit.—On adding a solution of permanganate of potash to a sewage, the permanganate becomes decolourized on account of its giving off a portion of its oxygen to the organic matter. Unfortunately, besides organic matter, salts of iron and other substances decolourize permanganate, so that however suitable the process is for the purpose of water analysis—for which Dr. Tidy originally introduced it—more reliance is to be placed upon the albuminoid ammonia as an index of the amount of organic matter present in sewage, particularly where iron salts are used as a precipitant. Speaking generally, the oxygen absorbed in four hours at 80° Fahr. is about ten times the amount of albuminoid ammonia.

The Mersey and Irwell Joint Board have adopted this process as a rough standard of purity for sewage effluents.

Taken in conjunction with the albuminoid ammonia process, reliable results can be obtained.

Standards of Purity for Sewage for Effluents.—No standard of purity for sewage effluents is fixed by the Rivers Pollution Prevention Acts, but from time to time various bodies have made suggestions as to what in their opinion should be adopted as a standard.

It will be admitted at once that any standard to be generally applicable, while being sufficiently high to prevent any nuisance arising from the contamination of a stream with an effluent which complies with it, should be sufficiently low to enable any Authority to keep within its limits. The Commissioners appointed in 1868 to inquire into the best means of preventing pollution of rivers recommended that it should be inadmissible to admit into any stream a liquid with the following composition:—

	Parts of 100,000.
Total suspended matter	4·0
Organic ,,	1·0
Organic carbon	2·0
Organic nitrogen	0·3

They then went on to give other conditions which apply only to the effluents from manufacturing processes. It is sufficient to say here that, in addition to the above, a sewage effluent should neither contain free alkali nor free acid.

The Local Government Board have more than once been asked to specify a standard of purity for sewage effluents, but they have refused, on the ground that all the circumstances of each case should be taken into consideration. In the practical working of the Rivers Pollution Prevention Act it is absolutely necessary that those who have to enforce the Act should have some standard before them. The Mersey and Irwell Joint Board have adopted 1·4 parts per 100,000 (one grain per gallon) of oxygen, absorbed at 80° Fahr. in four hours, as the standard for passable effluents; effluents absorbing less than 1 part being classed as good, less than 1·4 as fair, up to 3 as unsatisfactory, and more than 3 as bad. It is worth noting that in a recent report to the Board by their chief inspector it is shown that considerably over half the samples examined were within the limits of passable effluents. The standard which the Derbyshire County Council have adopted is 0·1 part per 100,000 of albuminoid ammonia. With regard to this standard, it can easily be reached by any of the recognized methods of treatment. The usual objection to the adoption of a general standard is that sanitary authorities would be tempted to dilute their effluent with subsoil or other water until the standard was reached. This would be prevented by insisting upon the effluent having undergone sufficient nitrification to contain at least 0·25 part of nitrogen as nitrates in the 100,000.

The following standard, taking four factors into consideration, would, in the author's opinion, be generally suitable :—

	Parts per 100,000.
Total suspended matter	less than 3·0
Oxygen absorbed at 80° Fahr. in 4 hours	,, ,, 1·5
Albuminoid ammonia	,, ,, 0·15
Nitrogen as nitrates to be at least	0·25

The Changes to be effected by Purification.—The typical average sewage which we have previously considered had the following composition expressed in parts per 100,000 :—

Total solids	142·0
Solids in suspension	42·0
Chlorine	12·0
Free ammonia	5·0
Organic ammonia	1·0

Such a sewage by precipitation would have the solid matter in suspension removed, and in this process the organic ammonia, "the index of the organic matter," would be reduced from 1 to perhaps 0·40. The same result would happen if the sewage was passed over the surface of land which intercepted the solid matter in suspension; the purification which would so far be carried out would be merely clarification or removal of suspended matter. The sewage would still have the organic matter in solution in it, and whether it is intermittently passed through the open soil, such as sand or gravel, or whether it has passed through artificially prepared filters of polarite, crushed coal, coke-breeze, coarse sand, or any other suitable substance, the change which takes place is the same.

It is essential that the sewage should be applied *intermittently* to the land or filters, so that the air in the interstices of the filtering medium may by this means be periodically renewed. Unless this takes place the nitrifying organisms cannot oxidize the ammoniacal salts and convert them first of all into nitrous acid, and finally into nitric acid, from two-thirds to three-fourths of which acids are oxygen. When these acids are formed they immediately replace the carbonic acid in the carbonates of lime, potash, or whatever base the carbonic acid present is combined with, and form nitrates of such bases. The nitrogen in solution therefore passes away in the sewage effluent as a nitrate; it is not removed. These nitrates are quite harmless; as a rule nitrate of lime is the particular nitrate which is formed. The general characteristics of nitrates are similar to nitrate of potash or saltpetre.

If such a sewage as the average one previously referred to were strained through filter-paper, the organic ammonia would be reduced from 1·0 to about 0·50 part per 100,000. All the matters in suspension would be removed, two-thirds of which we have seen are organic matter, being half the total organic matter in the sewage. In the laboratory, also, by carefully regulated experiments with precipitants, the whole of the suspended matters can be thrown down, and from 10 to 25 per cent. of the organic matter in solution; the organic ammonia, upon analysis, being reduced to 0·40, or slightly less. But in practice, by precipitation, the solid matters in suspension, and not more than 10 per cent. of the organic matter in solution, are removed.

The removal of the suspended matter is the easiest part of the process of sewage purification. The change is a physical one, whether the suspended matters are arrested mechanically on the surface of the soil, as in irrigation, or are entangled by a chemical during precipitation, and carried to the bottom of a precipitation tank. The rest of the purification is a biological action, the organic matter in solution not being removed, but oxidized by the nitrifying bacteria.

The results of the analysis of the average sewage, after going through these two changes, are exemplified in the following Table:—

	Total Solids.	Chlorine.	Parts per 100,000. Free Ammonia.	Organic Ammonia.	Nitrogen as Nitrites and Nitrates.
Raw sewage	142	12·0	5·0	1·0	Nil.
After precipitation	105	12·0	4·5	0·45	Nil.
Ditto and filtration	100	12·0	0·5	0·08	1·5

It will be noticed that by the precipitation the suspended matter has been removed, and that the solid matter in solution has practically only been diminished by the removal of the solid matter in suspension, while the chlorine has not diminished at all. An effluent, such as that above, if bottled up and kept in the dark at a warm temperature may become foul, but when diluted with running water and exposed to the air and light will not become a nuisance. The author has within his knowledge a case where a river, after receiving the effluent from sewage works, passed down crevices in the bed of the stream into a cavity in the limestone, where it stagnated without the access of air and light, and periodically overflowed into the river. At the times when this matter overflowed a considerable nuisance from sulphuretted hydrogen was occasioned. The organic ammonia in the mixed sewage effluent and river water upon analysis was found to be 0·10 part per 100,000; the effluent itself varied from 0·12 to 0·25 part per 100,000, and contained no nitrates.

Bottles of this mixture of river water and sewage effluent, in which mixture the organic ammonia was 0·11 part per 100,000, were hermetically sealed and placed in the dark for several weeks, when they gave off the smell of rotten eggs. The whole of the oxygen in solution in the water was taken up by the sewage matter. The above experiment is sufficient to show that the standard suggested of 0·1 per 100,000 of albuminoid ammonia is not unnecessarily stringent; the author suggested it to his Authority as a provisional standard until all the sanitary authorities in his county have levelled up to it. He is fully aware that many authorities turn out effluents with albuminoid ammonia less than 0·05 or even 0·04 per 100,000; but, in the first instance, it is advisable to fix such a standard that all Authorities without exception can comply with it.

The Rivers Pollution Commissioners, in their sixth report, gave a large number of analyses of effluents from sewage farms; the organic ammonia is not given, but the organic nitrogen is given as 0·191. If we take the organic ammonia which will be yielded as two-sevenths of this, 0·054 would be the organic ammonia in the effluents tabulated by them. It will be remembered that the organic nitrogen which they suggested should be taken as a standard in 1868 was 0·3; again, taking two-sevenths of this as the organic ammonia, we have a standard of 0·086, a more severe standard than the one suggested. The London County Council experiments on various methods of filtering sewage also showed that a standard of 0·1 per 100,000 could easily be complied with. The average result of 14 experiments on filtering through coke breeze was 0·1 per 100,000; and after filtering through polarite, 0·088 per 100,000.

The Sewage Committee of the British Association reported in 1870 that the effluent water from the sewage farm at Romford contained 0·037, 0·035, and 0·036 parts of albuminoid ammonia per 100,000. The following year (1871) they reported the effluent contained 0·064 and 0·045, while the effluent from Norwood farm contained 0·06, and the Tunbridge Wells farm 0·09; the effluent water from Reigate farm contained 0·06 part per 100,000 of albuminoid or organic ammonia; and the effluent from the Berley sewage farm contained from 0·039 to 0·05 parts per 100,000. It will, I think, therefore be conceded that a standard of 0·15 part per 100,000 of albuminoid ammonia, and 1·5 of oxygen absorbed at 80° Fahr. in four hours, is one that all Authorities should be called upon to comply with, the danger of diluting down to this standard being prevented by also insisting upon the effluent containing at least 0·25 part per 100,000 of nitrogen in the form of nitrates.

CHAPTER III.

VARIETIES OF SEWAGE AND THE CHANGES IT UNDERGOES.

Chemical composition of different sewages—Variations in the Chlorine—Night and day sewage—Hourly variations in flow and composition—Changes sewage undergoes by keeping—Purification effected by clarification, nitrification and bacteriolysis—Importance of sewers being self-cleansing—Table giving maximum discharge of various sized sewers at different inclinations.

THE following Table gives the average composition of a number of sewages. It will be seen that American sewage is much weaker than English, this of course being due to the larger amount of water used. Buxton also has an exceptionally weak sewage, the water from the celebrated public springs being passed into the sewers. It should be pointed out that Buxton is a water-closet town. A comparison of this sewage or that of London and of the 36 water-closet towns, with that from the 20 midden towns, clearly shows that the sewage from a water-closet town is no stronger than that from a midden town.

Average Composition of Sewage. Parts per 100,000.

Source of Sewage.	Total solids.	Solids in suspension.		Chlorine.	Ammonia.		Authority.
					Free.	Organic.	
American Sewage.							
(1) Laurence, Massachusetts	43·0	8·2		4·8	1·2	0·40	Reports of State Board of Health, Massachusetts.
(2) Worcester, Massachusetts	43·4	9·9 { organic mineral	4·7 5·2	4·3	0·55	0·178	Ditto.
English Sewage.							
(3) Buxton	45·0	5·5 { organic mineral	3·8 1·7	2·3	1·05	0·291	Author.
(4) Exeter	54·4	24·5 { organic mineral	14·5 10·0	5·0	3·7	0·212	W. J. Dibdin.
(5) London	123·5	39·6 { organic mineral	21·4 18·2	15·2	4·6	0·55	W. J. Dibdin.
(6) 20 midden towns	119·5	39·1 { organic mineral	21·3 17·8	11·5	5·4	0·56*	First Report of River Pollution Commissioners.
(7) 36 water-closet towns	116·9	44·7 { organic mineral	20·5 24·1	10·6	6·7	0·63*	Ditto.
(8) Salford	150·0	28·4 { organic mineral	16·4 12·0	41·0	1·84	0·68	Average of analysis by Dr. Burghead and Mr. Carter Bell.
(9) Chesterfield	130·0	54·0 { organic mineral	30·0 24·0	11·5	3·3	1·8	Author.
(10) Derby	137·0	57·0 { organic mineral	30·0 27·0	9·9	7·2	1·12	Author.
(11) Alfreton (few water-closets)	204·0	80·5 { organic mineral	48·0 33·5	17·4	8·3	1·64	Author.
(12) Sewage from slop-closet village, Stafford	158·0	...		19·6	19·7	2·58	Dr. George Reid.
(13) Wolverhampton (not an average sample)	496·0 { 110 114	50·0 { organic mineral	45·0 20·0	173·0	1·72	1·17	C. W. T. Jones, F.I.C.
(14) Burton-on-Trent	224 {	65·0		12·0	1·6	2·0	Author.
Ditto, after straining	...	...		...	...	0·88	
Ditto, after precipitation with lime and exposure to the air	...	...		...	1·8	0·68	
(15) Berlin	218·3	113·8 { organic mineral	75·5 38·3	21·8	12·8		Report on Berlin Sewage Farm.

* Estimated by taking ⅔ of the organic nitrogen.

The chlorine, it will be seen, varies from 2 to 173 parts per 100,000. In the strongest domestic sewage, when the water supply is only 5 or 6 gallons per head, the chlorine will be about 15 parts per 100,000 over that in the drinking water. The strongest domestic sewage the author is aware of is that given in sample No. 12, and the chlorine was 19·6. It was from a group of new houses, all of which were provided with slop water-closets, the excreta being removed by the slop water accumulating in tipplers which automatically flushed the closets.

In the case of the Berlin sewage the chlorine is 21·8. This high amount of chlorine is accounted for by the fact that from the baths in the Admiral's Gardens 2,000,000 kilograms of salt pass yearly into the sewers. At Salford the high chlorine is accounted for by the bleach, dye, and other manufacturers' wastes which enter the sewers; and the extremely high figure in the Wolverhampton sewage is caused, of course, by the galvanizers' pickle. The albuminoid or organic ammonia is on the whole the most important result of analysis to look at, as it is an index of the polluting organic matters. It varies from about 0·25 in weak sewage to 3, or occasionally more, parts per 100,000.

It has been pointed out that the composition of sewage varies at different times of the day. Thus the Sewage Committee of the British Association, in their Report for 1870, give the following analyses of Bury sewage :—

DAY AND NIGHT SEWAGE.

	Parts per 100,000.	
	Collected during day.	Collected during night.
Total solids	137	38·6
Solids in suspension	64·8	Trace
Chlorine	13·6	4·7
Ammonia	3·8	0·76
Organic ammonia	0·198	0·04

This variation in the composition of sewage is a matter that must be carefully borne in mind, as it has frequently been taken advantage of by interested persons to show an undue percentage of purification. If it takes five hours for the sewage to pass through the purification works, and a sample of the effluent and the raw sewage are taken at the same time, say about ten a.m., it will be evident that the effluent is from the night sewage; and in the case of Bury, though no purification at all were taking place, it might be claimed that the organic matter as indicated by the organic ammonia was being reduced nearly 80 per cent. The determination of the chlorine in the sewage would, however, show that the sewage and the effluent were not comparable. One reason for the night sewage becoming so weak is that water-taps are left running, as well as the dilution of the sewage with the subsoil water. This leakage of the subsoil water into the sewers took place to a much greater extent with the old brick sewers than it does with sanitary pipe sewers laid with cement joints.

In 1870 Mr. Bailey Denton reported to the Sewage Committee of the British Association that at Blackpool the sewage was actually 1,000,000 gallons a day, when the amount calculated from the water supply was 250,000 gallons, while at Hartford the discharge from the sewers was nine times the water supply. On a sewage farm recently constructed in the author's district, before a single house was coupled with the sewers, 40,000 gallons a day of water were delivered at the farm.

The following hourly estimations, by Mr. F. Perkins, F.I.C., of the chlorine in the Exeter sewage, show, when taken in conjunction with the hourly gaugings of the sewage, that its strength gradually increases in the daytime *pari passu* with the volume of the sewage :—

	Parts per 100,000. Chlorine.
6·20 a.m.	2·2
7·20 ,,	2·6
8·20 ,,	3·2
9·20 ,,	4·4
10·20 ,,	9·2
11·20 ,,	10·8
12·20 p.m.	10·2
1·20 ,,	8·0
2·20 ,,	8·7
3·20 ,,	6·0
4·20 ,,	14·25
5·20 ,,	7·0
6·20 ,,	7·0
7·20 ,,	5·2
8·20 ,,	5·0
9·20 ,,	6·0
10·20 ,,	4·4

	Parts per 100,000. Chlorine.
11·20 p.m.	4·6
12·20 a.m.	4·0
1·20 ,,	3·4
2·20 ,,	2·8
3·20 ,,	2·6
4·20 ,,	2·6
5·20 ,,	2·3

N.B.—The chlorine in the drinking water is 1·3, but that in the subsoil water will probably be higher.

In every case the sewage becomes weaker in the small hours of the morning, and where the sewers have tapped much ground water it becomes for an hour or two, about 4 a.m., quite clear and almost innocuous.

The diagrams in Fig. 2 (next page), after Santo Crimp, show the variation in the hourly flow of the sewage at a number of towns, and they should be compared with the estimations of the chlorine in the Exeter sewage.

As a rule, half the sewage passes off in eight hours, viz. 9 a.m. to 5 p.m. As there is always some leakage of subsoil water and some dilution from water-taps left running, it will be obvious that the sewage will be most diluted when the flow is least, and strongest when the flow is greatest. The fresher the sewage the less the organic matter in solution, and the more the sewage is agitated by pumping, etc., the more the solid matter is broken up and the more it enters into solution.

In addition to the variations due to the *amount* of water, a considerable difference in the nature of

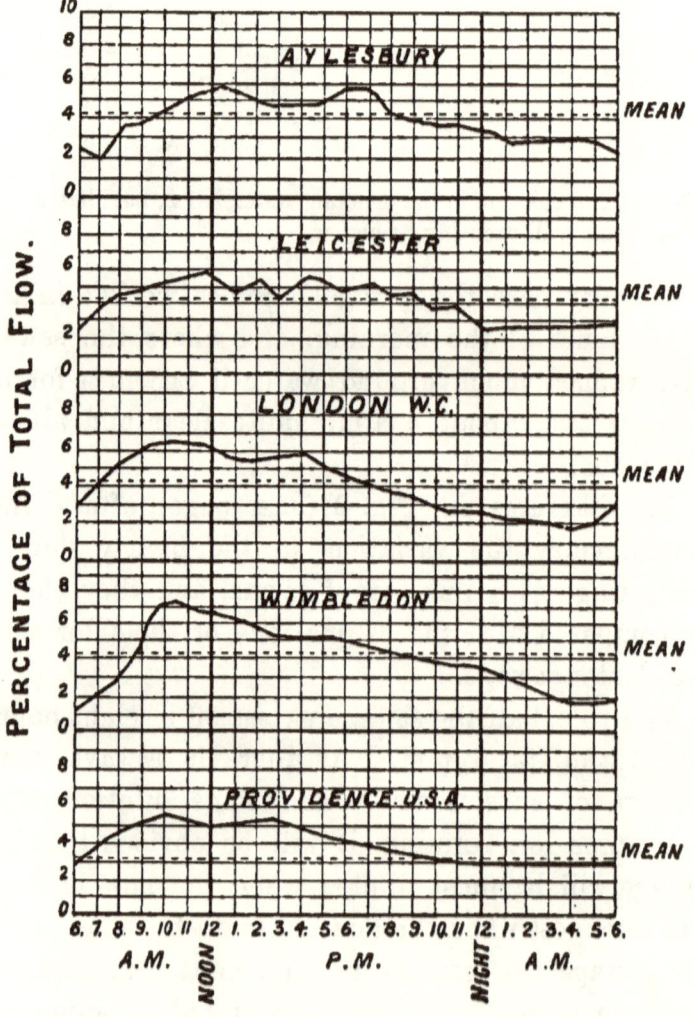

Fig. 2.—Diagram showing Hourly Flow of Sewage (reproduced, by permission, from W. Santo Crimp's "Sewage Disposal Works," published by C. Griffin & Co., Limited, London).

sewage is caused by the quality of the water in use in the district, and this is a point which has some influence on the question of treatment. From the point of view of the sewages they produce, waters may be grouped into three classes: soft waters, such as upland surface waters; waters with a moderate temporary hardness; and waters of a high permanent hardness containing sulphates.

The sewage from a town with a soft water will probably have to have lime added to it in the process of treatment to act as a base to the nitric acid which, we shall see, it is the object of sewage purification to form; while sewages formed from waters of the last group are liable to undergo decomposition with the production of sulphuretted hydrogen from the reduction of the sulphates; so that on the whole a water with a moderate temporary hardness from carbonate of lime in it forms the easiest sewage to treat.

From what has been said, it will be seen that sewage is an extremely complex and unstable liquid. When quite fresh it is practically inodorous, and, if not coloured by manufacturing wastes, is opalescent and of a light grey colour. Allowed to stand in a glass, a varying amount of solid matter will be seen to settle to the bottom, while particles of suspended matter, such as filaments from paper and fabrics, albuminous flocculi of animal tissues, and shreds of vegetable matter, will be found floating in it; and upon the surface particles of a fatty and soapy nature will accumulate.

The great bulk of the nitrogen in the sewage leaves the human body in the urine in the form of urea. Chemically, urea, which is the last product of the oxidation of albuminous foods, is of the following composition, $CO\begin{cases}NH_2\\NH_2\end{cases}$, and when obtained pure and dry assumes the form of colourless glistening crystals, without smell, and of a cooling and nitre-like taste. But almost immediately upon leaving the body it is attacked by one or other of two micro-organisms constantly present in the air, the *bacillus ureæ* and the *micrococcus ureæ* of Pasteur. Both of these organisms rapidly convert urea into carbonate of ammonia. The change which takes place may be expressed chemically as follows:—

$$\underset{\text{UREA}}{CO\begin{cases}NH_2\\NH_2\end{cases}} + \underset{\text{WATER.}}{OH_2} = \underset{\text{Carbonate of Ammonia.}}{(NH_4)_2\,CO_3}$$

This is the substance which gives stale urine and stables their pungent smell.

By the time the sewage reaches the sewage disposal works, as a rule the urea is completely transformed. If the sewage is kept without undergoing purification for a day or so it undergoes putrefaction, and begins to give off foul emanations; in the course of two or three days the albuminous matters begin to split up, and the sewage, particularly when the water

DECOMPOSITION OF SEWAGE.

contains sulphates, yields sulphuretted hydrogen, which is known by its characteristic odour of rotten eggs. When this gas is formed the sewage becomes black. As the above changes take place, more and more of the solid matter enters into solution, and the sewage becomes proportionately more difficult to treat, at any rate by a precipitation process.

The changes which take place may be expressed by the following empirical formula:—

$$2 \begin{cases} 7 \text{ atoms of carbon} \\ 1 \;\; ,, \;\; ,, \text{ hydrogen} \\ 2 \;\; ,, \;\; ,, \text{ nitrogen} \\ 3 \;\; ,, \;\; ,, \text{ oxygen} \end{cases} + 18 \begin{cases} 2 \text{ atoms of hydrogen} \\ 1 \;\; ,, \;\; ,, \text{ oxygen} \end{cases}$$

2 parts of albumen,* consisting of—   take up   18 parts of water, consisting of—

and thus form

$$2 \begin{cases} 1 \text{ atom of carbon} \\ 8 \text{ atoms of hydrogen} \\ 2 \;\; ,, \;\; ,, \text{ nitrogen} \\ 3 \;\; ,, \;\; ,, \text{ oxygen} \end{cases} \; 2 \begin{cases} 1 \text{ atom of carbon} \\ 4 \text{ atoms of hydrogen} \end{cases} \; 8 \begin{cases} 1 \text{ atom of carbon} \\ 2 \text{ atoms of oxygen} \end{cases}$$

2 parts of carbonate of ammonia, consisting of—   2 parts of marsh gas, consisting of—   8 parts of carbonic acid, consisting of—   7 parts of hydrogen consisting of 14 atoms of hydrogen

and 2 parts of carbon monoxide, consisting of 2 atoms of carbon and 2 atoms of oxygen;

or briefly thus:—

$$2C_7H N_2 O_3 + 18H_2O = 2(NH_4)_2CO_3 + 2CH_4 + 8CO_2 + 7H_2 + 2CO$$

From what has been said, I think it will be seen

---

* The percentage composition of albumen is approximately 52 parts carbon, 7 parts hydrogen, 22 parts oxygen, 16 nitrogen, and 1 sulphur.

that there can be no one process equally suitable for every sewage, but that the method of treatment for any particular sewage must depend upon the nature of the sewage, its freshness at the outfall, the character of the manufactures, and the quantity and quality of the water supply of the district. There are, however, certain general principles which are the same in every process of sewage purification, from irrigation on a sewage farm to the latest artificial schemes of purification by precipitation and filtration through specially constructed filters.

Instead of describing the various processes which are now in use according to the name of the process or the name of the inventor of the process, I propose to deal with the subject under three chief headings: (1) The defœcation or clarification of the sewage, by which I mean the removal of the solid matters in suspension, this generally being effected by precipitation; (2) The nitrification of sewage, by which is meant the conversion of the polluting nitrogenous matter in solution into nitrates; and (3) The bacterial process, or the new departure, including the Exeter septic process, in which the solid matters are split up by anaerobic germs which work in the absence of air, and the other processes in which aerobic organisms are utilized.

As the solid matter in suspension in the sewage can be entirely removed, while that in solution cannot be removed, but must be oxidized to render

DECOMPOSITION OF SEWAGE.

it harmless, it follows that the sewers should have a good gradient so as to bring the sewage as rapidly as possible to the outfall works. Where much manufacturing waste containing organic matter and hot water is poured into the sewers, there is still greater need for the sewage to be brought quickly to the outfall works. This is more especially the case with brewery waste, which contains much unstable organic matter, degenerate yeast, a lot of water used in the cooling, and, as a rule, a water supply containing sulphates.

When such conditions as the above are present, it is most important that there should be no chance of even a temporary deposit in the sewers, and every means should be taken to cause the sewage to flow as rapidly as possible, as unless this is done the organic matter will be split up, the sulphates reduced, and large volumes of sulphuretted hydrogen be evolved, which gas will enter into chemical combination with the iron salts or lime used as a precipitant, and thus render them to that extent inert, so that large quantities of the precipitant have to be used.

The following observations on "bottom velocities" were made by the late Mr. Beardmore:—

```
30 feet a minute will not move clay with sand and stones.
40   ,,        ,,   will move coarse sand.
60   ,,        ,,     ,,    ,,  pea gravel.
120  ,,        ,,     ,,    ,,  1 in. rounded pebbles.
180  ,,        ,,     ,,    ,,  1¾ in. angular stones.
```

The bottom is between three and four-fifths of the surface velocity, and it should always be sufficient to render the sewer self-cleansing. Under no condition should the velocity be less than two feet per second.

The following Table gives the maximum discharge per minute of various sized sewers within the range of permissible fall. A glance at it is sufficient to show that, if a sewer had only to convey the sewage undiluted with rain water, much smaller sewers than are generally used would be sufficient, but there is also a certain amount of rainfall which must be allowed to enter the sewers; this is usually calculated at $\frac{1}{4}$ of an inch per diem upon the inhabited area. After the normal dry weather sewage is diluted 8 times with rain water. It should be arranged for storm overflows discharging upon the surface of rough open filters to come into action.

TABLE GIVING THE MAXIMUM DISCHARGE*
IN GALLONS PER MINUTE OF VARIOUS SIZED PIPES AT DIFFERENT
INCLINATIONS WITHIN PERMISSIBLE LIMITS.

RATE OF INCLINATION.	4" PIPES. Discharge in gallons per min.	6" PIPES. Discharge in gallons per min.	9" PIPES. Discharge in gallons per min.	12" PIPES. Discharge in gallons per min.	15" PIPES. Discharge in gallons per min.	18" PIPES. Discharge in gallons per min.	24" PIPES. Discharge in gallons per min.
1 in 5 ...	410	1096	2902	5746	9728	14900	29059
1 ,, 6 ...	378	1016	2700	5369	9117	14009	27433
1 ,, 7 ...	352	951	2538	5061	8612	13267	26080
1 ,, 8 ...	332	898	2400	4801	8188	12632	24909
1 ,, 9 ...	314	851	2287	4581	7821	12077	23887
1 ,, 10 ...	299	812	2186	4385	7503	11603	22989
1 ,, 15 ...	246	674	1828	3695	6356	9865	19718
1 ,, 20 ...	210	588	1602	3250	5582	8743	17543
1 ,, 25 ...	191	528	1441	2936	5077	7927	15978
1 ,, 30 ...	174	483	1324	2697	4672	7310	14782
1 ,, 35 ...	161	446	1227	2510	4351	6817	13803
1 ,, 40 ...	151	418	1151	2354	4090	6406	12999
1 ,, 45 ...	141	393	1087	2222	3869	6060	12339
1 ,, 50 ...	133	374	1032	2113	3678	5781	11647
1 ,, 60 ...	122	340	944	1932	3364	5296	10787
1 ,, 70 ...	...	313	873	1790	3119	4911	10045
1 ,, 80 ...	...	291	814	1673	2920	4602	9416
1 ,, 90 ...	...	274	768	1575	2752	4338	8893
1 ,, 100 ...	...	260	727	1496	2614	4118	8456
1 ,, 120 ...	...	...	661	1359	2385	3765	7733
1 ,, 140 ...	...	...	607	1256	2202	3479	7148
1 ,, 160 ...	...	...	567	1174	2057	3259	6699
1 ,, 180 ...	...	...	...	1100	1934	3061	6306
1 ,, 200 ...	...	...	...	1041	1827	2896	5962
1 ,, 250 ...	...	...	...	924	1629	2576	5325
1 ,, 300 ...	...	...	...	841	1475	2345	4855
1 ,, 350 ...	...	...	...	...	1361	2158	4460
1 ,, 400 ...	...	...	...	...	1269	2015	4170
1 ,, 450 ...	...	...	...	...	...	1893	3916
1 ,, 500 ...	...	...	...	...	...	1784	3700
1 ,, 550 ...	...	...	...	...	...	1695	3524
1 ,, 600 ...	...	...	...	...	...	1618	3367
1 ,, 700 ...	...	...	...	...	...	...	3075
1 ,, 800 ...	...	...	...	...	...	...	2878
1 ,, 900 ...	...	...	...	...	...	...	2702
1 ,, 1000 ...	...	...	...	...	...	...	2545

* By Weisbach's formula: see also the "Engineer's Year Book." Crosby Lockwood & Son.

CHAPTER IV.

RIVER POLLUTION AND ITS EFFECTS.

Self-purification of rivers—Bacterial purification, by sedimentation, by biological changes—Gross pollution causing nuisance—Ptomain poisoning from polluted waters—Effects of polluted water on oyster culture—Typhoid fever and polluted river water.

River Pollution.—Having described what sewage is, and the various points that have an influence upon its quality and quantity, it would be well to briefly consider the results of pouring it directly into the streams and watercourses.

The Sixth Report of the Rivers Pollution Commissioners deals fully with the so-called self-purification of rivers. The Commissioners came to the conclusion that sewage mixed with twenty times its volume of pure water would only be about two-thirds purified in a flow of 168 miles, at the rate of one mile an hour. They also showed that a 5 per cent. solution of London sewage in highly oxygenated water was, in the course of five days, practically devoid of oxygen, and would therefore be unable to support fish life. It is probable that the Commissioners underestimated the amount of purification which takes place when the pollution is not sufficient to deprive the water of all its oxygen in solution. When the pollution is sufficiently great to deprive the water of all its oxygen in solution, the natural biological cycle of vegetable and animal water life is stopped, and no purification, except by subsidence and decomposition of mud, takes place.

When sewage matter is turned into a running stream the heavier insoluble mineral particles gradually subside, and in falling to the bottom they entangle a certain amount of organic matter, and some purification is thus effected. It is obvious, however, that where the current is not rapid, as in tide-locked rivers, in lakes and mill dams, a certain amount of silting up must occur. The extent to which this takes place will be appreciated when we examine the cost of dredging the Manchester Ship Canal, periodically published in the *Manchester Guardian*. I find that in 1896, 600,000 cubic yards of sludge were removed; 400,000 being contributed by the Irwell, and 200,000 by the Mersey and Bollin.

The nature of the deposit will be seen from the following analysis by Mr. W. Naylor, F.C.S., of the mud taken from Salford dock after being dried at 100° C. :—

Organic matter per cent.	Silica and insoluble matter per cent.	Carbonate of lime per cent.	Oxide of iron per cent.	Alumina and other metallic oxides per cent.
14	77·5	3·6	4·0	0·9

The river Rhone presents an excellent example of self-purification by subsidence. It enters the Lake of Geneva at its head, discolouring its waters for several miles by its rapid, turbid stream. A delta between six and seven miles long has been formed, and is said to be growing at the rate of a mile every five hundred years. Having thus deposited its mud, it leaves the lake a clear blue limpid stream.

Besides subsidence, there is a certain amount of oxidation taking place, particularly in the presence of salts of iron. Putrescent sewage reduces these salts to the ferrous state; ferrous sulphide giving sewage its characteristic black colour. These ferrous salts, in the presence of oxygen dissolved in water, become oxygenated ferric salts once more, to give up their oxygen to the sewage matter and again become ferrous salts. It is probable that nitrates are reduced to nitrites, and nitrites again to ammonia, through the action of putrescent organic matter in the presence of iron salts. Dr. Stevenson also suggests that compounds of iodine act as oxygen carriers.

Leaving the physical and chemical changes which take place when the sewage is added to running water, we now come to the consideration of the biological changes. The bacteriological life in river waters has during the last ten years been made the subject of many investigations in England and America, and on the Continent. The Water Research Committee of the Royal Society, and Dr. Frankland, have published in this country most important contributions upon this subject, more particularly upon the vitality of pathogenic organisms in river water. It is found that the typhoid bacillus and the bacillus of tuberculosis will live in river water for many days, the cholera organism for several months, and the anthrax bacillus for as long as two months.

One point upon which all authorities are agreed is that the number of bacteria in river waters in the summer months is much smaller than during the winter; the reason for this is that during the dry weather the rivers are chiefly formed of spring water, while in the wet season they receive the washings from manured fields, the sewage coming through storm overflows, and the washings from streets and cultivated land.

The bacteria in sewage may amount to almost any number, as many as 30,000,000 to a cubic centimetre having been estimated. From this it will be obvious that one of the effects of turning sewage into a river must be to increase its bacterial contents. The following are the results obtained in specific cases by well-known observers:—

THE DERWENT.

Above Derby.	Below Derby.
1870.	19440.

THE RHONE.

Above Lyons.	Below Lyons.
75 per c.c.	800 per c.c.

G. Roux.

The Saone.

586 per c.c.   4280 per c.c.

G. Roux.

The Spree.

In Berlin.   Below Berlin.
2000 to 20,000 per c.c.   50,000 to 500,000 per c.c.

At Sacrow below Havel Lake.
1500 to 20,000 per c.c.   M. Frank.

It should be explained that Havel Lake is seven miles long, and from a quarter to over one mile in width. The comparative great bacterial purity of the river at Sacrow is due to the subsidence and sedimentation in Havel Lake. Prausnitz gives the number of bacteria in the Iser above Munich as 530 per cubic centimetre, while 13 kilometres below the town it is 9111; 22 kilometres away the number falls to 4796; and at Freising, 33 kilometres below Munich, the number is reduced to 2378. The time occupied by the Iser in flowing from Munich to Freising is about eight hours.

Other remarkable instances of the bacterial purification of river waters by sedimentation are recorded by Dr. Percy Frankland;* but all the instances are open to the criticism that the results obtained are not wholly due to subsidence, but are also partly due to dilution. Some 150 varieties of bacilli and 40 varieties of micrococci in their growth attack organic matter and liquefy certain albuminous substances. The soluble products of the decomposition of this organic matter are absorbed by algæ and other plants

* Frankland, "Micro-Organisms in Water." Longmans.

of low type. These minute forms of vegetable life are the food stuff for rudimentary species of animal life, which serve in turn to nourish the larger animalculæ, which in their turn are the food stuff of fishes. These changes take place only when the water contains oxygen in solution. When the water contains no oxygen other series of bacteria grow in the water, and in their growth, as a rule, produce noxious emanations.

If the sewage is only small in amount, coarse fish live on the particles of fat, etc., which are suspended in it; eels and molluscs too flourish in the mud deposited, and effect a certain amount of purification. Here again, if the pollution goes beyond a certain limit, the fish will disappear, and the purification effected by them will cease. The green weeds, like all chlorophyllous plants, give off oxygen, which in the nascent state is a most active oxidizing agent.

It is not probable that nitrifying organisms grow in water to any great extent, as they require exposure to the air, but in mud the organic matter is attacked by putrefactive organisms, with the result that marsh gas is evolved; bubbles of this gas may be seen rising to the surface of any mud bank of a polluted stream, and in this way a certain amount of purification is effected. If the organic matter contains much sulphur, sulphuretted hydrogen is also evolved. This way of getting rid of the organic matter is most undesirable, as more or less of a nuisance is bound to be caused, and it is merely mentioned as a matter of scientific interest.

The nuisance formerly caused by the discharge of the unpurified sewage of London is notorious. The Metropolitan Sewage Commissioners reported in 1884 that in hot weather very serious nuisance was caused by the foul state of the water, the smell being most offensive, and the water unusable; that the sand dredged near the outfalls, which formerly was pure, was contaminated with foul mud, and that for fifteen miles below the outfalls the fish had disappeared from the river. During the Commissioners' inquiries they had occasion to embark upon the Thames, and three out of the five who went upon the river, together with the clerk who attended them, were attacked during the night of their excursion with severe diarrhœa, which they attributed to the nauseating odour from the river. Of course, nuisance of this kind arises where the sewage discharged into the river is large, and sufficient to use up all the available oxygen in solution in the water.

But it is not necessary for pollution to take place to this extent to become dangerous. In the grosser forms of pollution alluded to, the river water is, of course, unfit for use for any purposes except irrigation. It is not even fit for washing the floors, much less for milk-cooling, for cattle to drink, and the washing of dairy utensils. In this connection it is as well to point out that one of the commonest bacilli in ordinary sewage is the *bacillus coli communis*, the common bacillus of the animal intestine. In a report made to the London County Council by Dr. Andrewes and Mr. Lawes it was shown that the number of these bacilli present in sewage varied from 20,000 to 200,000 per cubic centimetre. As is well known, *bacilli coli* in their life processes elaborate dangerous ptomains, and if water in which these bacilli were present was used for washing out milk-vessels, disastrous consequences from ptomain poisoning might ensue. Similarly mussels from beds which receive sewage give rise to ptomain poisoning. When typhoid bacilli are in the sewage under the above circumstances outbreaks of typhoid fever would follow.

Where river water polluted with sewage is even more dangerous is where it has access to oyster-beds. The Twenty-fourth Annual Report of the Local Government Board contained a supplement on "Oyster Culture in relation to Disease." In this report it was clearly shown that a number of outbreaks of typhoid fever, both in this country and in America, were caused by the consumption of raw oysters, which had been exposed in the fattening-beds to sewage-contaminated water. Indeed, the report goes so far as to say that only a few of the fattening-beds in the country can be regarded as theoretically free from possible chance of sewage pollution.

As instances of the conditions then existing,

Cleethorpes may be quoted. These layings, which are among the largest in the country, are situated three-quarters of a mile below one outfall and a mile and a half below the other outfall from Cleethorpes, the two sewers bringing the sewage from a population of about 7500. From one mile to one mile and a half higher up the river than these sewers Grimsby, with a population of 60,000, drains into the river. Similar conditions exist at Southend, Medina river, and other well-known oyster-beds.

One of the well-known outbreaks of typhoid fever due to oysters is that which occurred amongst the students of Wesleyan University, Middletown, Connecticut, U.S.A. This outbreak affected a number of the students and their friends who attended the initiation suppers. The evidence in this case is as follows:—

1. The dates of the cases plainly point to a single source of infection.

2. This source of infection coincides as to date with the initiation suppers.

3. Four guests at the suppers, having no further connection with the College, developed typhoid simultaneously with the cases in the College.

4. The cases of typhoid occurred only amongst those attending three out of seven banquets.

5. Only one article of food was consumed at the three banquets that was not consumed at the other four, viz. raw oysters.

6. Twenty-five per cent. of those who consumed the oysters in question were attacked.

7. These oysters came from a creek 300 feet below the outfall of a sewer discharging sewage from a house where there was typhoid fever.

That the reality of this danger is appreciated by the public will be seen by the fact that the London Fish Trade Association petitioned the Government to ask Parliament to pass a Bill prohibiting the pollution of fisheries, and have invited the co-operation of the various County Councils in the matter.

Where the pollution is of a less gross nature, and the chemist has been unable even to detect it, outbreaks of typhoid fever have still been caused by drinking the water. The best known example of this kind is the now notorious outbreak of typhoid in the Tees Valley investigated by the late Dr. Barry, of the Local Government Board, of which Sir Richard Thorne has well said, " Seldom, if ever, has proof of the relation of water so befouled to wholesale occurrence of enteric fever been more obvious and patent."

The chief facts of this outbreak are shown on the diagram, Fig. 3. The population of the ten Registration Districts affected was 503,616. Of this number 219,435 drank water supplied by water companies who drew their supplies from the Tees, the water being filtered through sand-filters; the remaining population of 284,181 obtained their

TEES VALLEY OUTBREAK OF TYPHOID. 51

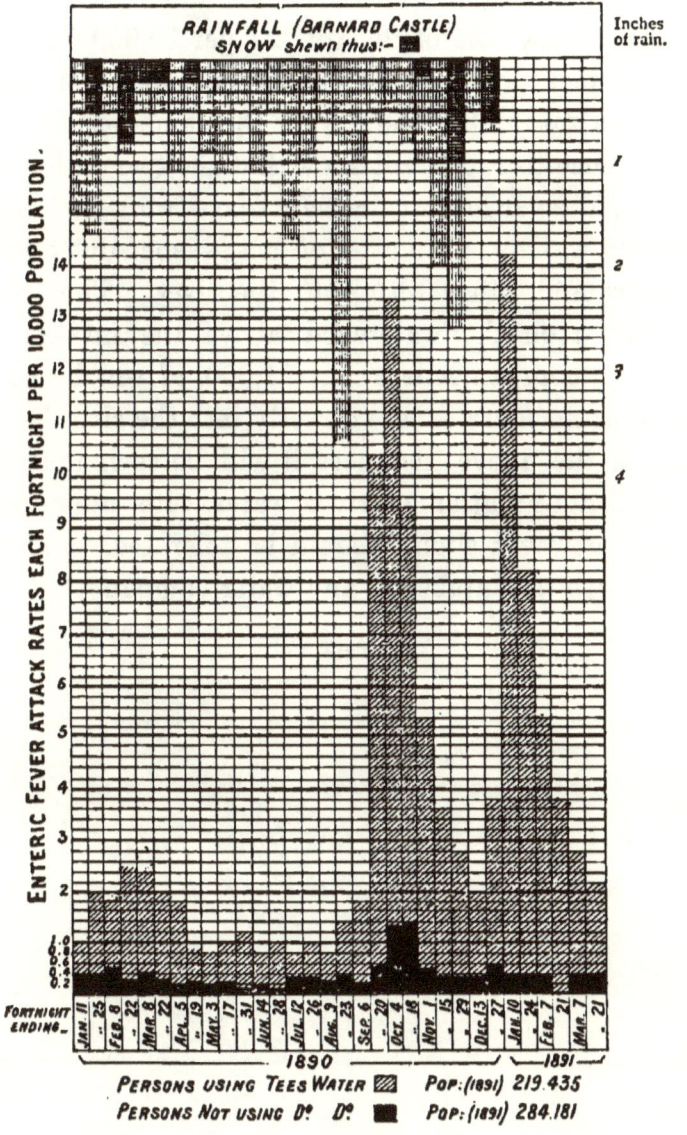

Fig. 3.—Diagram illustrative of OUTBREAK OF TYPHOID FEVER IN TEES VALLEY.

supplies from other sources. The first population had an attack rate of 29 and 24 per 10,000 for the two six-week periods, and the other population had an attack rate of 3·5 and 1·5 for the same periods.

Reference to Fig. 3 will show that the two outbursts of typhoid fever followed closely upon two very heavy rainfalls, which, as a matter of fact, washed out privy-middens and other filth into the river. The total population on the watershed above the intakes of the water companies is about 21,000, of which about 12,000 drain more or less directly into the river Tees.

From the above instances it will be seen that no direct contamination of river water with unpurified sewage, however slight, should be tolerated, particularly above the intakes of any water company.*

* Cf. page 121.

CHAPTER V.

THE LAND TREATMENT OF SEWAGE.

Sewage farms—Only indicated where soil is suitable—Clay lands nearly useless—The theory of nitrification—Intermittent downward filtration—Irrigation proper—Stratford-on-Avon farm—Edinburgh farm—Paris farm—Berlin farm—Proper method of laying out and working a farm—Quantity of land required—Fallacy of the manurial value of sewage.

Sewage Farms.—Perhaps the simplest method of purifying sewage, and undoubtedly the cheapest and the best where local circumstances permit of its being carried out, is by means of irrigation, by which I mean land treatment alone, including a certain amount of intermittent filtration. Unfortunately, however, it is not everywhere that a suitable soil is to be found, and if it can be found the price may be prohibitive. Nevertheless, the first inquiry to be undertaken with reference to the means of purifying the sewage of any district is to ascertain the nature of the soil and subsoil of the locality. For this purpose the author has found a short boring tool most useful. By means of a rod about four or five feet in length and half to three-quarters of an inch in diameter, made like a cheese-taster, pieces of core within three or four feet of the surface may be obtained. The geological maps are also extremely useful, but their accuracy must not be taken for granted. The best soil for the purpose of purifying sewage by irrigation is an open sandy soil, such as is met with in the Bunter sandstone, or the sandy gravels deposited in many valleys.

The older Authorities complicated the question as to what is the best soil for purifying sewage by trying at the same time to grow a remunerative crop. There can be no doubt that the answer given by the great bulk of the older Authorities to this particular question was the correct one—namely, a sandy or loamy soil, the soil not being of too open a nature. In fact, the soil should be of such porosity that it would absorb at the same rate that it would purify—that is to say, it should not be capable of absorbing more than 25,000 gallons per acre per diem, even if the sewage has the suspended matters first removed by means of precipitation.

But if the purification alone of sewage is aimed at, and no return from crops is looked for, a much more open soil is most suitable; and if the sewage, after undergoing a rough form of precipitation, is only applied intermittently for a few hours and the soil is given a rest for several days before any more is applied to it, a much larger quantity of sewage than this—indeed, as much as 100,000 gallons per diem per acre—can be purified.*

* Cf. actual working of Stratford-on-Avon intermittent filters, p. 69.

The danger with open soils is that more sewage will be applied than can be oxidized. What is generally taken by the average sewage farm as an indication not to apply more sewage is the fact that the soil begins to absorb less readily. It cannot be too emphatically stated that when this point is reached the land is *overdosed*. Provided, however, that too much sewage is not applied to the land, and if care is taken in this respect, the more open the soil is the better, if it is a good thickness and the subsoil is fine gravel and uniform in texture. Even blown sand on the seacoast will thoroughly purify sewage, if it is applied in small quantities at a time and the application is stopped before the sewage reaches half-way to the land-drains.

It cannot be too clearly understood that there is no relation between the quantity of sewage that can be got to pass through the soil and the quantity of sewage that can be purified; for instance, a thin soil overlying a shaly open rock will take any amount of sewage, but will not purify more than a retentive clay.

The following are some of the formations the soils of which are suitable for purifying sewage: alluvial drift and gravel, the chalk, oolitic sandstones, bunter sandstone, and the magnesian limestone when sufficiently weathered; the old red sandstone, and occasionally the millstone grit.

In almost every formation beds of clay are to be found, and it should not suffice to have one or two trial-holes sunk, but trial-holes should be dug on every side of the piece of land it is proposed to irrigate, and bore-holes to a depth of four feet should be made at intervals all over the site.

The most unsuitable soil, and, unfortunately, one of our commonest, is clay land. It is said that clay lands can be rendered more fit for filtration by ploughing and digging-in ashes, which convert the impervious surface and allow the sewage to sink through. There are in Derbyshire two farms upon which considerable sums of money have been spent in thus preparing the land, in one instance as much as £1123 being spent in lightening 14 acres of land to a depth of two feet with engine ashes. It is perfectly true that this enables the sewage to pass through the clay, but it does not lead to the purification of the sewage, and where the land is a stiff clay it undoubtedly would be better to construct sewage filters.

Clay lands, besides being too impermeable to permit the sewage to pass through them, are unfortunately open to another objection—viz. that in dry weather they crack and fissure, so that the sewage passes directly through the cracks to the land-drains without undergoing any purification. Worms also leave permanent holes in clay, which last for a considerable length of time, and permit the sewage to pass down.

At a small farm at Brampton in Derbyshire the sewage contained a considerable amount of dye-water, and upon a trial-hole being sunk on the said farm, which is a stiff clay, the author found innumerable worm-holes passing directly downwards to the effluent drains, the worm-holes having their sides saturated with dye, and showing how the sewage passed away absolutely unpurified.

Instead of attempting to lighten clay lands where this is the only land available, sewage filters should undoubtedly be adopted.

The researches of Warington, Schloesing, and Muntz, and of the Massachusetts State Board of Health, have shown that the number of nitrifying organisms in soils of sewage filters rapidly diminishes from the surface downwards. Small quantities of soil taken at different depths from the surface at Rothamsted showed that in clay soils no nitrification takes place at a greater depth than eighteen inches; in the most porous soils, however, nitrification still took place at a depth of four feet from the surface. When we bear in mind the quantity of oxygen* necessary for the growth of the nitrifying organisms, these researches only confirm what one would suppose to be the case from general reasoning. It is possible, however, by under-draining very open soils to a great depth, that air may

* About half the weight of the nitrates produced consists of oxygen.

be taken down deeper and the process be carried on even below four feet.

Fig. 4 shows the number of bacteria per gram of sand at varying depths in a sewage filter, as published in the returns of the State Board of Health, Massachusetts. Reference to the diagram—which is taken from a communication to the Institute of Civil Engineers by C. H. Cooper, C.E.—will show that in the first quarter of an inch the bacteria amounted to over 1,100,000, and at five inches from the surface the number had dropped to about 100,000, so that it will be seen that the great bulk of the changes effected take place in its upper layers.

The change effected by a sewage-filter is identical with that effected by the soil, and reference should now be made to the chapter on Nitrification. The only difference is that when sewage is filtered through land an attempt is generally made to utilize the nitrates that are formed by the nitrifying organisms for the growth of vegetables, grass, osiers, etc. It is only necessary to point out now that the vegetables do not absorb the nitrogen until it is first converted into nitrates, or, at any rate, into nitrites, by the nitrifying organisms.

It will lead to a clearer understanding of the proper method of applying sewage to land to briefly state the method which is usually adopted in filtering sewage. The sewage is applied to the sewage filter for six or eight hours, a penstock at the outlet

NITRIFICATION CHIEFLY AT SURFACE.

at the bottom of the filter being partially closed so that the sewage passes slowly and uniformly through the filter, in order that every particle of sewage shall

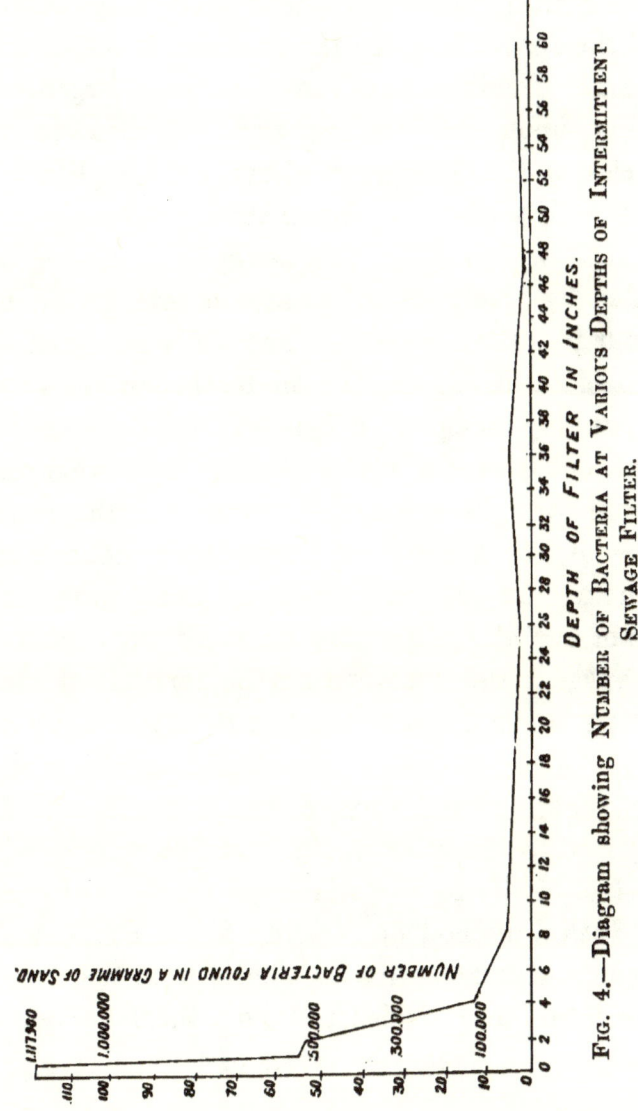

FIG. 4.—Diagram showing Number of Bacteria at Various Depths of Intermittent Sewage Filter.

be brought into intimate contact with the filtering medium. After six hours the penstock is opened wide, and all the sewage is allowed to drain out.

As the filter drains dry, air is drawn into the filter to fill the spaces between the particles of the filtering medium. In this way the nitrifying organisms, growing upon the surface and attached to the particles of the filtering medium, are supplied with oxygen. The whole arrangement is frequently made automatic by allowing the sewage (of course first clarified by precipitation) to accumulate during half an hour in a tank, which automatically discharges by means of a flushing syphon on to the surface of the filter, and by means of a penstock regulate the rate of flow out from the filter, so that the sewage does not run straight through the filter, but the surface drains dry to a depth of about three inches every half-hour. In this way air is drawn into the interstices of the filter supplying the nitrifying organisms with oxygen, and surrounding the particles of which the filter is composed with alternate layers of air and sewage. After the end of the half-hour the tank is again filled, and the syphon discharges the clarified sewage on to the filter again. In this way the vital process of nitrifying bacteria are enabled to be carried on. Such a method of aërating filters could not be carried on unless ventilating shafts from the bottom of the filter are provided to let out the layers of air carried down between the layers of sewage.

INTERMITTENT FILTRATION.

Of course if sewage was applied to land every half-hour through day and night all the year round no crops could be grown; but if the soil was open the land would no doubt purify the sewage quite as well as an artificial sewage filter.

No hard-and-fast line can be drawn between irrigation and intermittent filtration. The change it is desired to effect is brought about by nitrifying bacteria, and it is the same in each case, except that in irrigation crops are grown on the area irrigated. The sewage being applied with a view of manuring the crops, the surface of the soil is laid out with a uniform fall, so that the sewage will trickle over the surface in a thin layer. In intermittent filtration the surface of the soil is first levelled and then laid out in ditches about eighteen inches wide at the bottom and three feet at the top, the spaces

FIG. 5.—METHOD OF PREPARING LAND FOR INTERMITTENT DOWNWARD FILTRATION.—Section on ⅛th scale.

between the ditches being two or sometimes three feet in width, as shown in Fig. 5. In this case crops are grown on the ridges and the ditch is used exactly as a sewage filter.

Intermittent filtration is intensified irrigation. It is, in fact, a compromise between irrigation and the adoption of sewage filters. It is true that crops are grown on the ridges between the ditches, the sewage percolating laterally to their roots. On a properly laid out sewage farm there should be part of the land prepared for irrigation and part for intermittent downward filtration, and it is a good plan to alternate the land which is used for these two purposes, as on the Berlin, Birmingham, and other large sewage farms.

Irrigation Proper.—The usual method of carrying out irrigation is for the sewage to be brought to the highest level of the surrounding land, generally through an underground pipe; from this point it is conveyed through an open earthenware or concrete carrier along a contour; branch distributing carriers being given off in its course. These branch distributing carriers may be allowed to overflow on to the surface of the soil which should be laid out for a uniform fall of about 1 in 100. The actual fall should depend upon the impermeability of the soil. With impermeable soil the fall being even less than this. The final distributing carriers may be merely furrows in the soil itself. With pervious soils far more distributing carriers are necessary to spread the sewage

SURFACE IRRIGATION. 63

evenly over the land than are necessary with retentive soils.

With regard to stiff clay lands, the best results will be obtained by irrigating twice over, or even more times, as at Leicester, by a catch-water system, as in Fig. 5.

SECTION ON LINE A.B.

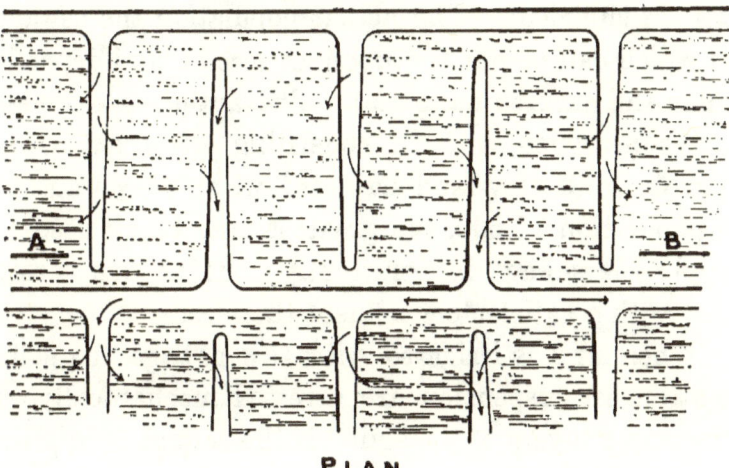

PLAN

FIG. 6.—METHOD OF IRRIGATING STIFF CLAY LAND.

The arrangement of the distributing carriers must to some extent depend upon the conformation of the land. If it has a gentle slope or is irregular a catch-water system should be adopted, the distributing carriers running along the slope at various levels, the usual distance allowed between two carriers being about a chain; but this distance must depend purely upon the slope and the permeability of the soil, with a rapid slope it being a greater distance, and with a very pervious soil less. When the ground is level, it must be prepared in broad ridges on the ridge and furrow system, the beds being about forty feet wide with the feeder along the top of the ridge.

If the sewage is first clarified by precipitation, from half to one-third of the land otherwise necessary will be sufficient. The actual population the sewage of which can be purified upon an acre of land depends upon the nature of the soil. The author has under his supervision a farm consisting of about forty acres of stiff clay land irrigated with the sewage of a population of 600, the greatest care being used to apply the sewage intermittently; but he has rarely met with a satisfactory effluent therefrom.

When land is irrigated with raw sewage the solid matters in suspension are arrested mechanically on the upper layers of the soil. If the sewage is applied for longer than three or four hours to the same land, the surface nearest to the distributing carrier becomes covered with a layer of slime, which, when the sewage ceases to be applied to the land, dries and forms an impervious film. This not only kills the crops, but also prevents the air from entering the soil and stops nitrification. Where irrigation is adopted, the system used at Berlin—of passing the

SCREENING ARRANGEMENTS.

Fig. 7.—Cage Sewage Screens, as used at Buxton.

PRELIMINARY TREATMENT OF SEWAGE. 67

sewage through roughing tanks to allow deposition of the grosser suspended matters—should be taken advantage of; or the sewage should be passed through a series of cage screens, such as are shown in Fig. 7, which arrest a much larger proportion of solids than the ordinary fixed screens.

Unless the land is of a very porous and open nature it will be necessary to do even more than this, either by constructing precipitating tanks and adopting at least partial precipitation, or, what will have the same effect, adding a precipitant to the sewage and allowing it to pass through large carriers in which the suspended matter will be deposited, before applying it to the land.

At Burton-on-Trent sewage farm some five years ago no precipitant was adopted, the result being that the solids in suspension coated the surface of the soil, preventing the sewage from passing in, so that it stood about in big lagoons all over the farm. These lagoons of sewage underwent decomposition, and damages on account of the nuisance caused thereby were obtained by a gentleman who experienced the nuisance at his house five miles away from the farm. As the result of this and other actions brought against the corporation, lime has been adopted as a precipitant. No precipitation tanks have, however, been constructed, the sewage merely running through carriers in the ground for about one hundred yards. In this run, the solid matters in suspension become deposited, and the sewage is at the end of the run perfectly clear; and the soil absorbs it quite readily.

The same experience was obtained at Stratford-on-Avon, where after an injunction had been obtained against the corporation a precipitant was added to the sewage. In this case, too, no precipitating tanks were constructed. After the use of a precipitant it was found that the surface of the land no longer became occluded, and, as a consequence, the sewage no longer lies about in sickening pools creating a nuisance. The following is the method adopted of applying the sewage to the land at Stratford-on-Avon.

The population of Stratford is about 9000, and the volume of sewage about 270,000 gallons a day. At night it is stored in a tank, and pumped on to the land during ten hours each day. Although some fourteen acres of stiff clay land is occasionally irrigated, practically the whole of the sewage is disposed of on nine acres of open sandy gravel, the whole of which has been levelled and is underdrained (the drains being six yards to twelve yards apart); half of it is used for irrigation (so-called), and half is laid out in ridges three feet wide, and furrows one foot wide and nine inches deep, for intermittent filtration. Every year the area in ridges and furrows is levelled, and the area irrigated is converted into ridges and furrows; in this way the land gets well worked up and aërated. Italian rye-grass is grown upon the flat beds, and mangolds and cabbage on the ridges, the sewage only being applied to the furrows. Not a little of the success of the scheme is due to the intermittent manner in which the sewage is applied.

The Stratford effluent contained only 0·056 part of albuminoid ammonia per 100,000, coming well below the provisional standard of 0·100 part which the author has suggested. That nitrification was actually taking place, in spite of an intense frost upon the occasion of the author's visit, is proved by the fact that the effluent contained 1·5 parts of nitrogen per 100,000 in the form of nitrates and nitrites.

The level plot is divided into eight beds, each of which is temporarily divided into four quarters by earthen ridges. The sewage is applied to each quarter intermittently for twenty hours; the bed then has a rest for about six weeks, no sewage being applied. The following is given as the actual time that the sewage was applied to a particular quarter of the bed:—

	Hrs. Mins.	Hrs. Mins.
Sunday 8.40 to 9.50 a.m. ...	1 10	
„ 11.30 to 12.0 a.m. ...	0 30	
		1 40
Monday 8.20 to 9.30 a.m. ...	1 10	
„ 12.15 to 1.0 p.m. ...	45	
„ 2.15 to 3.0 p.m. ...	45	
„ 4.30 to 5.0 p.m. ...	30	
„ 6.20 to 7.45 p.m. ...	1 25	
		4 35

	Hrs. Mins.	Hrs. Mins.
Brought forward		4 35
Tuesday, Wednesday, and Thursday, as on Monday, each	4 35	
		13 45
Total number of hours sewage applied		20 0

After this treatment, no more sewage is put on for six weeks. The sewage delivered through the feed-pipes is 22,680 gallons per hour, so that in the 20 hours 453,600 gallons would be applied to one-eighth part of an acre in six weeks.

By applying the sewage in this intermittent manner for a few hours at a time, air is drawn into the soil between the layers of the sewage (in fact, at Stratford-on-Avon, air could be seen bubbling out of the soil when some fresh sewage was being put on), and not only is the sewage thoroughly purified, but excellent crops are grown on the ridges.*

When a sewage farm is laid out, it should be laid out in plots of a convenient size, which should all be numbered, and a record should be kept in a book to show exactly what plots are being irrigated and the time that they are irrigated for. It is always advisable to have part of a farm laid out as an osier-bed for the reception of storm-water, and it is also advisable to have on the farm a specially prepared

* I have to express my indebtedness to Mr. A. H. Campbell, the borough engineer, for the above details respecting the method of working the farm.

sewage filter to take the sewage in times of rain, and at such times as the application of sewage to the crops would be injurious. The following diagram (Fig. 8) illustrates the proper method of laying out a sewage farm. A is the detritus tank, B the artificial filter, C the coarse filter for storm-water, D the intermittent land filters laid out in ridge and furrow, E the irrigation area, while the osier bed skirts the river.

With regard to the crops, Italian rye-grass is undoubtedly the best, as there seems to be no limit to its capacity for absorbing sewage; it also grows so closely and so rapidly as to choke down weeds, the seeds of which are brought to the farm in the sewage. Unfortunately, however, rye-grass has to be used freshly cut; it also should not be irrigated for at least ten days or a fortnight before being cut.

Where the sewage farm is large enough for dairy cows to be kept, the difficulty is to a great extent met, because the crops can be cut and consumed on the farm by the cows. Of course, the cows on a sewage farm are kept under cover all the year round. The Corporation of Birmingham have acted on this principle for many years with most satisfactory results. The grass plots should be ploughed up about every three years, and they should be re-sown every spring; as Italian rye-grass and Timothy grass stand the winter very badly.

The next most useful plant for irrigation, particularly

on clay and retentive soils, is osiers; and it should be borne in mind that the roots of osiers penetrate

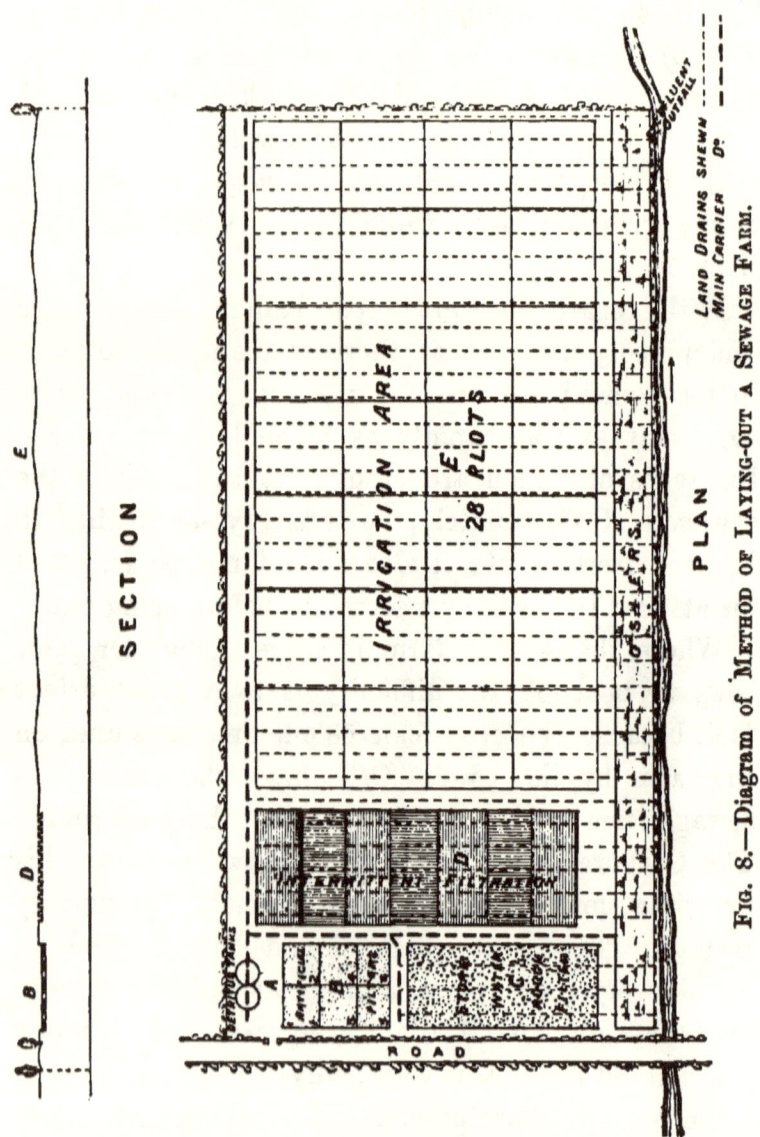

Fig. 8.—Diagram of Method of Laying-out a Sewage Farm.

to a great depth, so that it is of no use underdraining a site on which osiers are to be planted, as the roots soon choke the land drains.

The Local Government Board Committee on the Modes of Treating Town Sewage reported in 1876, that, after rye-grass, cow cabbage and mangolds were the only crops that would continuously take sewage and flourish.

The farm which has the most suitable soil for the purification of sewage within the author's knowledge is that of the Nottingham sewage farm. The Craigentinny sewage fields of Edinburgh are very often referred to, chiefly, perhaps, as instances where sewage disposal works have been made to show a profitable return. The soil here is blown sand, which, before it was irrigated, was perfectly barren; but these fields should never be quoted as sewage *purification* works, as the sewage is applied in great volumes, and it runs away practically unpurified into the drains. The following description of the Craigentinny fields, by Mr. W. Fairley, C.E., is taken from the *Proceedings of the Institute of Civil Engineers*:—

" On the Central or Craigentinny drainage area there are two irrigation farms, both situated on the north-east outskirts of the city, viz. the Lochend of the Craigentinny meadows. Lochend farm has an area of forty-three acres, of which thirty-five acres are commanded by main carriers led from the point at which the main outfall sewer from the city debouches into an open stream. The sewage is pumped up on to the remaining eight acres by an undershot water-wheel driven by the sewage flowing to the lower area. The Craigentinny irrigation farm extends to the foreshore of the Firth of Forth, along which it spreads for a distance of about 1½ miles. This farm, which has an area of about 236 acres, may be taken as typical of the others. The surface of the ground is undulating with a general slope seawards, and a large portion of it was formerly waste land. There is a complete system of under-drainage, consisting of two-inch and three-inch agricultural tile drains, laid ten yards apart and three feet deep. The effluent water of the surface and the under drains is caught by ditches, to be again used in irrigating lower levels, or delivered by the outlets into the Firth of Forth. The secondary carriers divide the ground into plots, each about three-quarters of an acre in area. Four crops are usually taken off the ground annually, and occasionally in dry and favourable years five or six. The ground was originally sown with a mixture of natural and aquatic grasses, the latter in small quantities. The fifty acres under Italian grass are situated at a higher level than the main area of the farm, and are cropped in rotation with the arable land. All these sewage farms are in the hands of private individuals, and they no doubt form a good investment, for after the meadows have been laid out the expenses entailed in the maintenance and management are very small. The grass, which is annually put up in lots and sold by public auction, is readily bought by cowkeepers and dairy-farmers. It cannot be said that these farms play an important part in the *purification* of the sewage. The great bulk of the foul water merely runs over the surface of the ground, and deposits a portion of its suspended matter; but where under-drainage has been provided, the effluent is fairly clear and pure. The presence of the meadows has a tendency to lower the value of land for building purposes in the immediate vicinity. Although no distinct nuisance may arise from them, their near neighbourhood is not free from disagreeable odours.

"The Central or Craigentinny Drainage District comprises an area of 1455 acres, with an estimated population of 107,000, the dry weather sewage discharge being about 800 cubic feet per minute.

"At Portobello, a watering place much frequented by the residents of Edinburgh, frequent complaints were made by the corporation of the pollution of the stream caused by the sewage of Edinburgh, and after litigation the referee recommended that an outfall sewer should be carried direct to the sea with a discharge of the water of Leith outfall.

"By an agreement with the proprietor of the Craigentinny estate, provision had to be made to intercept the sewage discharged on the foreshore from the central outlet of the city, and for this purpose a cast-iron pipe 2 feet 10 inches in diameter was laid across the beach joining the 3 feet 6 inches pipe at an angle of 45 degrees. At the main outlet the sewage is discharged into a strong current, the general set of which is N.E.; any floating matter is therefore carried out to sea, clear of Portobello, and the bathing beaches along the coast."

It must be admitted, however, that what is done at Craigentinny is no criterion inland, or even as to what should be done by an important corporation like that of Edinburgh, even though it has a tidal river to discharge its effluent into.

Another instance of a barren waste becoming a fertile tract of land by being irrigated with sewage is near Paris. The land here, again, is extremely porous and open. In this case, under the unanimous advice of the experts consulted, the Paris municipal authorities decided to dispose of their sewage by irrigation on a neck of land at Genne Villiers. Thirty years ago this land was a mere waste; it is now a fertile market garden. The Authorities have been so satisfied with the results obtained that twelve years ago further land was obtained at Achères. Altogether, Paris has an area of nearly 4000 acres available for the purification of the sewage of its two and a half million inhabitants. This instance is also no criterion for this country, as the rainfall and climate are both exceptionally favourable for sewage farming.

The largest sewage farm in the world is that at Berlin. In March, 1895, the available area for irrigation was 22,881 acres, while the population draining to it is about one and three-quarter millions. The subsoil is sand, with patches of loam and argillaceous loam. The price paid for the land was about £40 per acre. What has made it possible for this farm to be carried on on such a scale, and so well, is the fact that the labour is generally provided by convicts, so that the actual cost of labour on the farm cannot be accurately estimated. Everything connected with the whole of the arrangements is carried on in a manner that would be quite impossible in this country. The day starts with a parade of the sewage men. At six, every man is given his instructions for the day in writing, with a note what the punishment will be if he fails to carry them out. The whole farm is divided into plots of five or six acres in extent, surrounded by roads twenty feet wide and planted with fruit trees. Each plot is further subdivided into so-called "quarters," two-thirds of an acre in extent. The surface of the ground is most carefully levelled to an average slope of one in thirty-three, and upon this slope Timothy and Italian rye-grass are grown. The sewage is brought along the top of the slope, and allowed to flow uniformly over it in a thin layer down the slope. The grass plots are ploughed every four or five years. Before the sewage is allowed to flow over the surface of the slope, it is passed through roughing tanks, or catch pits, as it has been found that the coating of sludge on the land prevents the access of air to the soil, and indeed kills the grass.

The information given above has been extracted from several papers which have been read before the Sanitary Institute and the Institute of Civil Engineers by Mr. Roechling, C.E., who also gives the following details relating to the Berlin sewage farms :—

No. of acres of total area to million gallons of sewage.	Acres actually treated for each million gallons of sewage.	Persons to each acre actually treated.	Persons to each acre of farm.	Gallons of sewage per acre.
372	268	156	112	2687

A large portion of the land is also laid out as intermittent downward filtration beds in ridge and furrow, the ridges being three feet wide and the furrows or ditches eighteen inches deep and two feet six inches wide (see Fig. 5). The sewage is allowed to flow into the ditches until they are nearly full, but not sufficiently high to get on to the top of the ridges so as to soil the turnips, mangolds, and cabbages, which are grown upon them. The solid matter in the sewage falls to the bottom of the ditch or furrow, and impedes the soaking of the sewage through it, the sewage therefore passes laterally to the roots of the plants. The farm has been thoroughly drained on the parallel system in contradistinction to the herring-bone principle, the average depth of the under-drain being three to four feet. The total quantity of sewage being thirty million gallons per day.

The proper method of working a sewage farm is to have about one-eighth of it laid out in ditches for downward intermittent filtration (see Fig. 5, page 61). The site of the intermittent filter should be changed each year, so that the whole of the farm is systematically broken up. The area used for intermittent filtration should be divided into twenty-eight beds, four for each day of the week. Boards, with the names of the days upon which the particular beds are to receive the sewage, should be placed against them, so that any one visiting the farm can at once see that the manager is applying the sewage in accordance with his instructions. The crops grown should be absolutely subservient to the question of purification. The rest of the farm being used for irrigation, the requirements of the crops can, to a certain extent, be considered.

It is also advisable to have a filter capable, in an emergency, of taking the whole of the day's sewage. The filter may be made of crushed stone, coarse sand, or any material which will pass through a three-sixteenths of an inch mesh, and should be under-drained to a depth of four feet. If a sewage filter is constructed it should have a small quantity of sewage applied to it at least twice a week, otherwise it may become dry and the nitrifying organisms upon it perish. Such a filter will be of the very greatest value for use in wet weather, when the crops do not require the sewage, and it will also be useful when the crops are about to be cut. Of course, nothing at all should be grown upon the surface of the filter, not even weeds. The site of the sewage filter will, of course, always remain the same, but the intermittent filtration area should, as I have said, be changed each year, so that the same piece of land becomes an intermittent filter once every eight years. During the time it is being used for intermittent filtration it will store up a considerable amount of manure, which will be utilized while the land is being irrigated.

If the sewage farm is really to do the work required

of it, a steam plough is a useful adjunct, so as to break up the surface and to admit the air to a depth of two feet, and in laying out a farm a permanent haulage station, with movable system of cables, should also be devised. In addition, screening arrangements, such as those shown in Fig. 7, and a rough filter for storm water, are necessary.

The conditions under which irrigation alone should be adopted are where there is an open sandy loam or loamy gravel, which can be obtained at a price not much exceeding £150 an acre, and where there is no manufacturing refuse in the sewage which will either cause a nuisance or be detrimental to the crops. Where the land is of a suitable nature, but the price is prohibitive, from half to one-third of the area will suffice if precipitation is also adopted. Whether it is worth while to go to the extra annual expense of precipitation, depends entirely upon the price that is asked for the land. The following Table gives the author's opinion as to the varying quantities of land, with the system which is adopted:—

1. *Irrigation alone.*
    From 1 acre to every 25 persons with stiff clay.
    To   1   „   „   200   „   „   sandy gravel.
2. *Irrigation and Intermittent Filtration.*
    From 1 acre to every 200 persons with alluvial drift.
    To   1   „   „   400   „   „   sandy gravel.

The one great argument that has repeatedly been used in favour of irrigation, is that the nitrogen of the sewage is utilized by the growing of crops, and that we have in sewage a material source of wealth, the money value of which on theoretical calculations was fixed by the Rivers Pollution Commissioners at about 2*d.* per ton. The suspended matter in 100 tons of average sewage was said to be worth 2*s.*, while the matter in solution was worth 15*s*. Hoffman and Witt, and Lawes and Gilbert, arrived at similar conclusions, while Mr. Bailey Denton put down the value at 1¾*d.* per ton.

Such being the theoretical value of sewage, it was only natural that an attempt should be made to extract its valuable constituents; but as Dr. Tidy pointed out, in a most learned paper read before the Society of Arts, the utilization of sewage must be distinguished from its purification—they are totally different questions. It is true that sewage may be poured in enormous quantities upon crops growing upon an open soil, such as blown sand, and that under these circumstances a certain proportion of the valuable constituents of sewage would be utilized, but the sewage must be poured on in such quantities that little purification will be effected. When the nitrogen, as nitrates and nitrites in a good effluent, is estimated, it will be found that the crops have taken up a very small percentage of nitrogen indeed. In addition to this, it is doubtful whether all the nitrogen taken up is derived from the crops. The researches of Nobbe and Hiltner on leguminous plants have shown that by the aid of certain bacilli small tubercles are formed in the roots of the leguminous plants which enable them to take up nitrogen from the air.

In many of the cases which have been quoted as instances where the total nitrogen has been greatly diminished in passing sewage through land, and where it is assumed that the nitrogen has been absorbed by the crops, it will be seen that the chlorine has also diminished, the fact being that the sewage has become diluted with subsoil water. The analyses published by the British Association of the sewage and effluent from Breton's farm, Romford, the Croydon farm, and Merthyr Tydvil, are all open to this criticism. It also applies to Sir Edward Frankland's experiments on the Barking farm, as reported in 1871. In this case 70 per cent. of the nitrogen came away in the sewage; but it is wrong to assume, as has been done, that the other 30 per cent. of the nitrogen was absorbed by the crops, for the chlorine was diminished 34 per cent., an amount which certainly could not be absorbed by vegetables, and a figure which clearly shows that the effluent was greatly diluted with subsoil water.

It can be shown that there is a considerable amount of ammonia in smoke, and when urging upon manufacturers the necessity for using smoke-preventing appliances, this fact has ingeniously been adopted as an argument why black smoke should be allowed to continue or even be encouraged. The argument is quite as sound as that of the money value of sewage.

When the English reports as to the money value of sewage went across the Atlantic, Forbes replied to them by asking, if a bottle of brandy were poured into a barrel of water, whether the mixture would be worth as much as the original brandy; while Professor Storer pointed out that Philadelphia stands upon a bed of clay which contains a pound of gold in every 1,224,000 pounds. From which data, the money value of the gold in the clay within the limits of the city of Philadelphia is not less than 100,000,000 dollars, but no one would dream of attempting to extract it. These American criticisms are strictly to the point. Agriculturists can get chemical fertilizers in a form in which they can be readily used for less money than the labour costs to apply the sewage in a manner in which it can be absorbed by the crops.

We must not lose sight of the fact that the nitrogen in the sewage has to be converted into nitrates, or nitrites, before it is assimilable by the crops, and to do this it must be applied with equal volumes of air, and the cost of the labour in so applying it must be taken into the estimate. Liebig's fantastic picture of Britain as a vampire on Europe, sucking out its life-blood and pouring it into the sea, was conceived without taking into consideration the fact of the enormous quantities of nitrogen which annually are imported into the country in the form of beef, mutton, corn, and nitrates, nor did it take into consideration the enormous quantities of carbonate and sulphate of ammonia, annually manufactured in the country at the various gas-works, which are now used as fertilizers.

When we bear these facts in mind, and also that leguminous plants have the power of absorbing nitrogen from the air, and that the sea requires its nitrogen replenishing, on account of the fish taken from it, the fallacy of the manurial-value argument becomes too ridiculous to entertain for one moment.*

* Since the above was written, Prof. Crookes has endeavoured to resuscitate the nitrogen bogey, with the evident intention of showing how easily it could be laid low, by utilizing the nitrogen of the air as a source of nitrates.

CHAPTER VI.

PRECIPITATION, PRECIPITANTS, AND TANKS.

Limitations of the process—In practice solid matters only removed—Soluble matter removed by excess of precipitants—Comparative costs of precipitants—Lime, alum, copperas—Various patent precipitants—Precipitation without tanks—Absolute rest tanks—Continuous-flow tanks—Dortmund tank of Kinebühler—The Candy tank—Sludge, and methods of dealing with it.

By precipitation is meant the deposition of the insoluble matter in suspension in sewage, together with a certain proportion of the organic matter in solution, by the addition of some chemical which forms insoluble compounds with it, and at the same time deodorizes it. For a considerable time it was claimed by one school of experts that by precipitation sewage could be sufficiently purified to be discharged into any river, while a rival school advocated the claims of land alone. To-day the limitations of each are recognized, and it is generally admitted that by precipitation practically only the solid matter in suspension is removed.

The chemicals which have been used for this purpose are almost innumerable; but only three need seriously be considered, viz. alum, copperas, and lime, or some combination of these. Most of the precipitants sold under fanciful trade names are composed of these substances in varying proportions. The general plan of the process of precipitation is for the chemical to be added to the sewage; the mixed chemical and sewage is then well mixed up by means of water-wheels or baffling plates. The mixture is then allowed to flow into a large precipitation tank, where it is either allowed to rest for a few hours, or is passed continuously through the tank at such a slow rate of velocity that the suspended matters fall to the bottom of the tank, where they form a dark sludge, containing about 95 per cent. of water; the clarified effluent overflowing from the tank. Details of different forms of precipitation tanks are given on a subsequent page.

The use of lime as a precipitant has been much prejudiced by the reports which have been published on the so-called "lime process" as it was carried out years ago, the effluent from the precipitating tank merely being passed into the river without any irrigation or filtration. It should be at once clearly stated that the effluent obtained by any process of precipitation is not fit to be turned into a stream without subsequent biological filtration. For practical purposes, it may be assumed that the precipitation process will merely remove the solid matters in suspension. It is true, as Mr. Dibdin has shown, that in the laboratory from 10 to 30 per cent., and even more, of the organic matter *in solution* can be removed; but it is a safe rule to assume that in the working of a sewage purification scheme by precipitation the suspended matters only are removed.

The following Table gives the results obtained by Dibdin with various precipitants upon the organic matter *in solution* in the Metropolitan sewage. The results are expressed as a reduction per cent. in the oxidizable organic matter:—

REDUCTION OF SOLUBLE ORGANIC MATTER BY VARIOUS PRECIPITANTS.

Grains per gallon of precipitant used.			Percentage of purification.
3·7 of lime *in solution*			11
5·0 ,, ,,			15
10·0 ,, ,,			19
15·0 ,, ,,			25
3·7 ,, ,,	and 2·5 of iron sulphate		18
3·7 ,, ,,	5·0 ,, ,,		21
5·0 ,, ,,	10·0 ,, ,,		25
10·0 ,, ,,	10·0 ,, ,,		30
5·0 ,, ,,	5·0 of sulphate of alum		18
28·0 ,, ,,	20·0 ,, ,,		
	6·0 of sulphate of iron		24
56·0 ,, ,,	40 of sulphate of alum and 12 of sulphate of iron		31

In addition to Mr. Dibdin's experiments, the Massachusetts State Board of Health have made careful experiments on the relative value of various precipitants, and the proportion of organic matter *in solution* removable by them. The following Table gives in a condensed form some of the most practical of their conclusions, the quantities of the precipitants used being in each case of the same value, about 1*s.* 3*d.* per annum per head of the population dealt with :—

Crude sewage yielding 0·40 part per 100,000 of organic matter.
  ,,   after settling 0·28 ,,   ,,   ,,   ,,
  ,,   ,, straining 0·26 ,,   ,,   ,,   ,,

After precipitation with—

	Organic ammonia per 100,000.	Reduction of soluble matter per cent.
1800 lbs. of lime per million gallons	0·19	0·22
1000 lbs. of copperas and 700 lbs. of lime per million gallons	0·17	0·29
400 lbs. of ferric sulphate	0·15	0·32
650 lbs. of alum sulphate	0·19	0·20

With regard to lime, what takes place when it is added to sewage is that immediately carbonate of lime is formed; this acts as a weighting material, entangling the flocculent matters in suspension and carrying them down to the bottom of the tank with it. The lime also forms an insoluble compound with a certain amount of organic matter in solution, which also is carried down.

There is no doubt that the carbonic acid in the sewage holds a lot of carbonate of lime in solution, and if the tank effluent is allowed to pass into any stream with a short run the carbonic acid is given off, and the carbonate of lime is deposited as a white insoluble substance in the river. This is what is sometimes called the secondary decomposition of lime effluents. If, however, the tank effluent obtained from precipitation with lime is irrigated over a large area, the carbonate of lime is deposited in the soil, and is also split up by the nitric acid formed by the nitrifying organisms, a stable nitrate of lime being formed in its place, and no secondary decomposition will result.

Another objection which there was to the use of lime as a precipitant, without subsequent intermittent filtration, arises from the danger of the sewage being overdosed, so that some free lime might escape into the rivers and thus kill the fish. Wherever lime is used the following conditions should be complied with—

1. An indicator, such as phenol-phthalein or nitrate of silver, should be used, and the lime should only be added until the sewage becomes barely alkaline.

2. The lime should be ground and added as a saturated lime-water.

3. Mechanical agitation of the sewage with lime-water should immediately take place.

The sludge should be removed if possible *each day* from the precipitating tanks. It soon putrefies and rises in large pieces, setting up decomposition of the supernatant sewage, thus undoing the good done by chemical treatment.

There are three classes of sewage for which lime is particularly suitable.

1. Sewage containing salts of iron and mineral acids, such as the sewage from Birmingham, Sheffield, and Wolverhampton. With this class of sewage, the lime has the advantage that it neutralizes the acid which would otherwise absolutely prevent nitrification; it also combines with the iron salts and forms hydrated oxide of iron, which acts also as an excellent precipitant.

2. Sewage containing the refuse from breweries. Brewery waste contains a large amount of carbonic acid and yeast cells which produce carbonic acid. This with the lime forms the insoluble carbonate previously alluded to. The sewage of Burton-on-Trent is precipitated with lime, as much as thirty to forty grains to the gallon being used.

3. Sewage containing dye wastes. In some instances it may be necessary to also add a small proportion of copperas in this case.

With all precipitation processes the sewage should be got to the outfall works as soon as possible, so that as much organic matter as can shall be retained as suspended matter. For the same reason, when it is necessary to pump the sewage for the purpose of filtration, *this is best done after it has gone through the precipitation tanks.*

For *small* schemes it is practically impossible to use lime as a precipitant, on account of the necessity for expensive machinery to grind it. The commonest precipitant used in this country for small schemes

ALUMINO-FERRIC.

is alum. Spence's alumino-ferric blocks is a favourite method of employing it. These blocks are merely suspended in the sewage, and as the flow increases more of the precipitant comes in contact with the sewage and is dissolved, precipitating the fatty and albuminous matters to the bottom of the tank as sludge. This precipitant costs from £2 10s. to £3 per ton. The quantity necessary to precipitate the suspended matter is from 10 to 20 grains per gallon, and by reckoning $15\frac{1}{2}$ grains * per gallon as equivalent to a ton per million gallons, the cost can easily be calculated.

The various chemicals which have been suggested for use as precipitants are innumerable. Many of them are combinations of alum and iron salts and lime. The following is a list of some of the most known precipitants:—

Name of Precipitant.	Chief Ingredients.
Ferrozone	Crude alum, copperas, and magnetic oxide.
Alumino-ferric	Crude alum, with a trace of iron salts.
A. B. C.	Alum, blood, and clay. The alum and the clay are the useful ingredients. Probably as good results would be obtained by the alum alone.
Amines process	Lime and herring brine. Large quantities of lime are used, and the temporary sterilizing of the sewage is, no doubt, due to it.

* A ton actually contains 15,680,000 grains.

SEWAGE PURIFICATION.

Name of Precipitant.	Chief Ingredients.
Bacillite process	A disinfectant, such as carbolic acid, is used as well as a precipitant. The germicide, if real, would also kill the nitrifying organisms, so that this process, like the amines, should not be countenanced if it really is a sterilizing process.
Hanson's process	Black ash waste, consisting of salts of soda and sulphide and sulphite of lime. The precipitant will have to be oxidized before nitrification can start, and it is probable lime alone would be as efficient.
The phosphate process	Phosphate of alum in hydrochloric or sulphuric acid. The idea was that a triple phosphate would be precipitated, the ammonia thus being trapped, as it were. The value of the ammonia obtained in practice does not compensate for the phosphate lost.
The electrolysis process	The sewage is electrolysed, oxygen being liberated; this partly attacks the organic matter and partly the iron plates at the negative pool, forming iron salts, which act as a precipitant. The iron salts can be added cheaper as copperas in the ordinary way.
The Hermite process	Electrolysed sea-water, containing free chlorine, is added to the sewage, with the view of sterilizing it. (See remarks on Bacillite process.)

Precipitation without Tanks.—At Stratford-on-Avon and at Burton, the experiment is being tried of adding a precipitant without any precipitation tanks. As far as Stratford-on-Avon is concerned, the experiment is quite successful. The solids in suspension are thrown down upon the surface of the land and are dug in. At Burton-upon-Trent the same method is being adopted, but here the enormous quantity of sludge deposited is deteriorating the porosity of the land. The late Dr. Tidy was responsible for the Stratford-on-Avon experiment, while the Burton experiment (where thirty to forty grains of lime per gallon are employed) is due to Professor Dewar. Before the sewage was limed at Burton, the slimy sludge was carried *all over* the surface of the land, covering it with a sticky albuminous film. Since a precipitant has been used the sludge is deposited on a limited area, and the sewage at present passes freely through the land. But where a precipitant is used, and there is not a very large area of land, tanks of some kind or other must be constructed.

For the success of precipitation processes, besides the sewage being fresh and as little broken up as possible, it is necessary that the precipitants should be got into very intimate union with a small portion of the sewage, and that this should then be mixed carefully with the whole of it.

The tanks must be adequate in size. Their actual capacity in any case will depend upon whether they are to be used on the absolute rest principle or continuously; if for absolute rest for six or eight hours, the tanks must be in triplicate, and should hold at least a day's sewage.

94 SEWAGE PURIFICATION.

Absolute Rest Tanks.—These tanks can only be used without pumping where there is an available fall of six or eight feet for the precipitation portion of the process of purification. As a fall of six or eight inches only is required in continuous tanks, and any extra fall which is available can be utilized to greater advantage by filtering twice over, I do not think that absolute rest tanks will be used to any extent in the future.

Where there is such a great fall at the outfall works that absolute rest tanks are decided upon, the best arrangement is as sketched in Fig. 9; it will be seen that the clarified sewage runs out through a floating arm near the bottom of the tanks, thereby requiring a considerable fall.

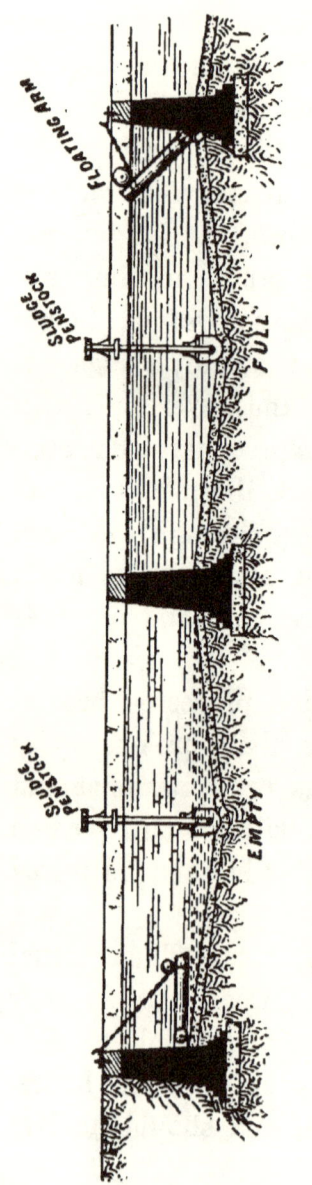

Fig. 9.—Complete-rest Precipitation Tanks.—Section on $\frac{1}{10}$th scale.

ABSOLUTE REST TANKS.

The illustration (Fig. 9) shows the ordinary complete rest precipitation tanks in the various stages of the process of purification. No. 1 is being emptied by means of a floating arm, the outlet of the floating arm is about four inches below the surface, and falls with the level of the sewage. The result is, that the sludge and the scum are both left in the tank, and only the clarified sewage is decanted. The sewage is commanded by a valve, which prevents it escaping through the floating arm. The sludge is let off by the sludge penstock being opened. In the process of filling, the sewage runs down a floating salmon ladder to prevent the sludge being stirred up.

Where there is not more fall than two or three feet, the precipitation must take place in a continuous-flow tank, and in this country, until the last few years, the arrangements for removing the sludge have been of a most unsatisfactory nature. *First of all it has been necessary to have the tanks in duplicate, and, under certain circumstances, in triplicate, then it has been necessary to pump off the tank effluent before the sludge could be removed.*

Continuous-flow Tanks.—The Conical-bottomed Dortmund Tank of Carl Kinebühler.—The credit for first getting over these difficulties is due to Carl Kinebühler, for although the tanks designed by him and adopted at Dortmund and subsequently at the Chicago Exhibition, and during the last three years, with slight modifications, at numerous places in England, have not proved absolutely satisfactory when only used singly, yet, compared with the alternative of the old-fashioned, flat-bottomed, continuous-flow tank, they possess advantages which make them immensely superior, and when constructed in duplicate, with arrangements for emptying them when necessary, they leave little or nothing to be desired.

The drawbacks to them are said to be—

1. The sludge does not always gravitate to the bottom of the cone.

2. As the sides are not thoroughly cleaned, colonies of bacteria form, and are carried off in the tank effluent.

With regard to the first objection, I have in my district about a dozen conical bottomed tanks, and I can emphatically state, if the sludge is removed daily, this difficulty does not arise. On the other hand, some of the most solid sludge I have ever met with has come from tanks of this description, and contained as much as 7 per cent. of solid matter.

With regard to the second objection, that, after a time, colonies of bacteria pass over in the tank effluent, so as to injure the filters, I must corroborate this; but it is a difficulty which could be easily overcome by the adoption of a revolving scraper, of a rapid screw shape, which, by being slowly revolved each day, at the time the sludge was being pumped out, would remove the slime from the sides of the tank, and wind it down to the apex of the cone. Another way of obviating this difficulty is to have at the outlet a series of perforated pipes four inches below the water level, across the tank, or to have the tanks in duplicate.

But whether the conical bottom survives in the process of the evolution of a perfect precipitating tank or not, the apparently paradoxical continuous upward flow of the sewage, with a downward movement of the sludge, is an innovation which has come to stop.

There is in these upward-flow tanks a neutral plane at the level of the point, where the velocity is so slow that the particles precipitated cannot be carried on. Each particle of the coagulum arrests another in its upward path until, growing as a rolling snowball does, there is across the whole tank a floating meshwork of flocculent matter, which acts as a kind of strainer.

That this is so, I have proved by watching the process in a small tank constructed of glass, and I have also frequently passed a glass tube down a precipitating tank, and, on drawing it up, have found the precipitation all confined within levels twelve to eighteen inches apart. The following diagram of the Dortmund tank clearly explains the method of working it:—

98 SEWAGE PURIFICATION.

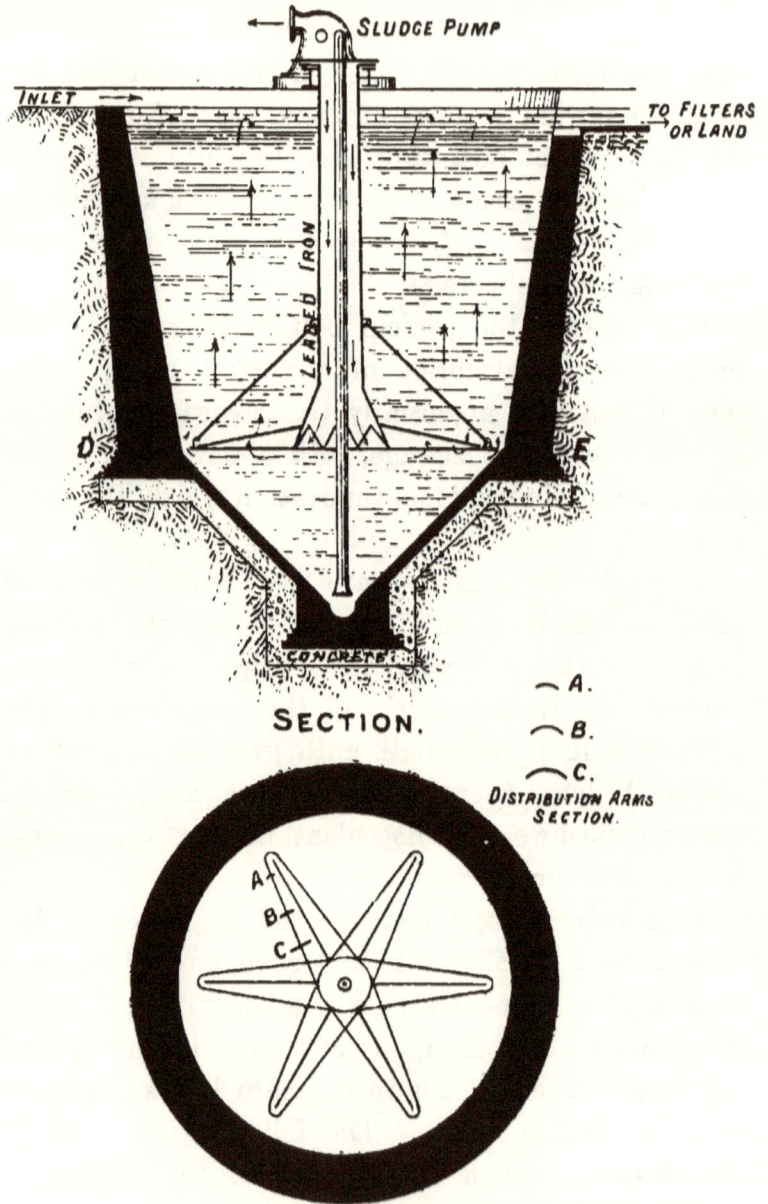

Fig. 10.—Improved Dortmund Tank.—Scale 8 feet to the inch.

The Candy Tank.—The upward-flow principle has been adopted by Mr. Candy in his circular flat-bottomed tank. This tank permits of the sludge being removed without first pumping off the supernatant tank effluent; this can be done by gravitation where there is a fall of about eighteen inches.

The author has two of these tanks in his district, and has carefully watched them for the last two years. They act admirably if the sludge is removed daily. The sludge, which, in his experience, contains about 97 per cent. of water, is removed by revolving a perforated sludge-pipe, which is pivoted at the centre of the tank floor, and by means of a worm gear sweeps all round the bottom of the tank; the outlet of the sludge-pipe is made to discharge eighteen inches below the level of the sewage in the tank, and when the screw-down valve is opened, the head of eighteen inches of water in the tank is sufficient to force the sludge through the perforations and up the sludge-pipe. When the sludge begins to run thin, the man in charge slowly winds the revolving perforated arm round, until one complete revolution has been made, by which time the whole of the sludge lying on the bottom of the tank will have been picked up. Attached to the revolving perforated sludge-pipe is a rubber scraper, which, by the same mechanism as moves the sludge-pipe, scrapes the sides and bottom of the tank, thus preventing the growth of colonies of bacteria, which is one of the drawbacks to the Dortmund tank.

The International Sewage Purification Company have also applied the same principle of extracting

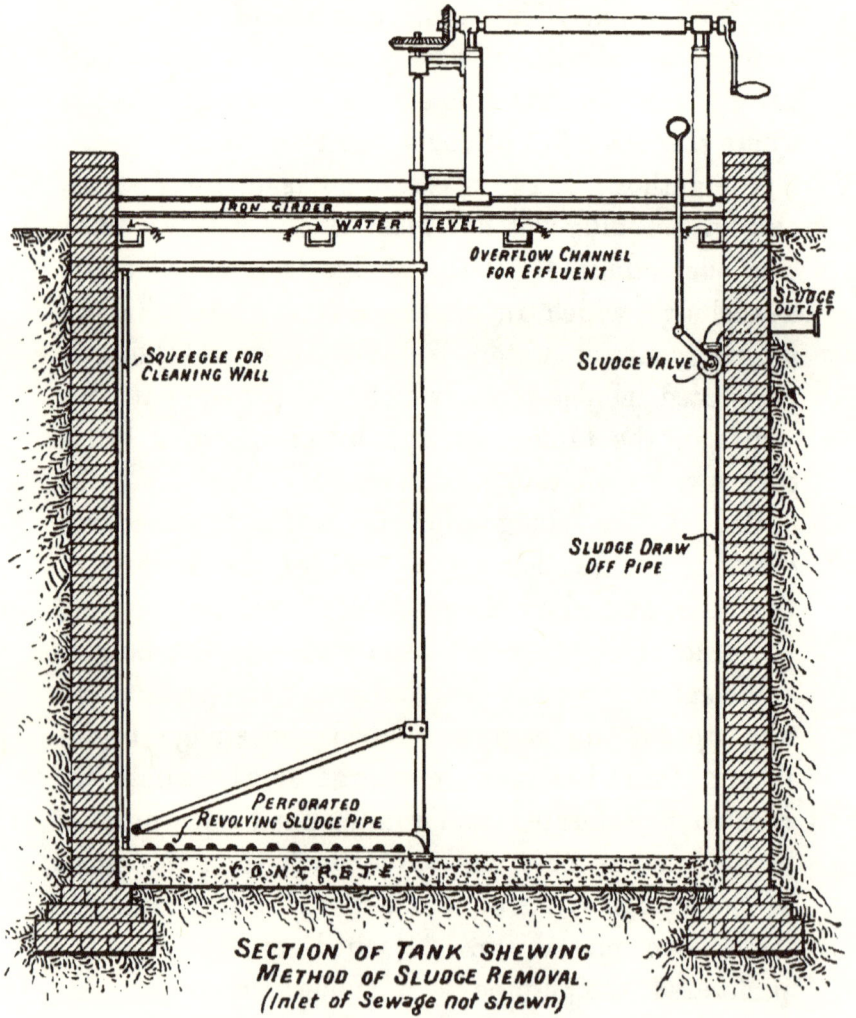

Fig. 11.—The Candy Tank.

sludge to rectangular tanks, the perforated arm travelling from one end to the other on a tramway.

Perhaps the latest form of tank which permits the sludge to be removed without first pumping off the tank effluent, is what may be called the "travelling squeegee" tank. In this, the sludge is forced down to the deep end of the ordinary rectangular continuous-flow precipitating tank, by means of a slowly moving squeegee, which travels on rails laid on the bottom of the tank. When the squeegee has been brought within about three yards of a perforated sludge-pipe, lying at the deepest part of the tank, the sludge outlet is opened, and the squeegee moved up gradually to this pipe, as the sludge flows out about eighteen inches below the top water level. The squeegee is then wound back to the top of the tank, where it stays till the sludge is removed, the next day.

There are several other precipitating tanks, such as the "Ives," a modification of the ordinary Dortmund tank, but not (to the author's mind) an improvement on it. There are others which need not be mentioned, as one has no word of approval for them.

With regard to the percentage of purification effected by precipitation, having taken the average of some twenty precipitation tanks in his district, the author has found that the purification effected by alumino-ferric on ordinary domestic sewage is about 60 per cent.

With regard to the precipitant to be adopted under various conditions, with ordinary domestic sewage, where there is a large proportion of soapy slop water, nothing could act better than alum. Spence's alumino-ferric cakes acting on such sewage in a Dortmund tank gave the following results :—

	Organic ammonia, parts per 100,000.
Before precipitation	0·88
After ,,	0·36

For sewage containing mineral acids or brewery wastes, lime is the most suitable precipitant. For sewage containing dye waste, copperas and lime, and for sewage containing tannin, copperas, alum, and lime will probably be required. The exact proportions of these various ingredients will vary with the particular sewage.

To show how commonly precipitation is now adopted in addition to irrigation, a return made to Parliament in 1894 showed that this process was adopted at as many as 174 places.

In sixty cases the return also gives the nature of the precipitants used, and from it I have constructed the following Table, showing the nature of the chief precipitants in use, and the number of places at which they were adopted :—

Lime.	Alumino-ferric.	Lime and alumino-ferric.	Ferro-zone.	A.B.C.	Alum.	Iron, salts, and lime.	Lime, alum, and copperas.
20	11	8	9	2	4	5	1

SLUDGE.

Sludge.—After the sludge has been removed from the precipitating tanks, in small schemes—that is to say, in schemes for populations of less than 5000, and where the outfall works are far removed from houses—it is sufficient to allow it to dry in trenches about one foot deep and one yard wide, the trenches being covered at the bottom with ashes, and having a drain leading back to the precipitating tanks. In this way the liquid part of the sludge is got rid of, and in a few weeks' time it dries into a friable cake, which makes an excellent top-dressing for grass, but is not of such value that it can be sold.

For larger schemes, special areas of land will have to be set apart for drying the sludge upon, and then movable carriers will have to be provided for turning it upon the different parts of the land. This method of removing the sludge has been successfully worked at Birmingham for many years, and it is distinctly indicated wherever there is any low-lying land, the level of which it is desirable to raise, and there are no houses within a few hundred yards of the sludge-drying area.

The sludge is also useful for embanking rivers to prevent storm waters from overflowing the land.

Another method of dealing with it very commonly in vogue, particularly where the works are close to habitations, is to press it with a filter press. This press consists of two folds of jute or other coarse fabric, spread between plates of iron, whose surfaces are grooved; the sludge is forced through the jute by means of compressed air, and the water is blown out, leaving the solids between the two layers of jute as a cake of sludge about one inch in thickness and a couple of feet square.

Unfortunately this method of dealing with the sludge is somewhat expensive, the cost of the process working out at about 2s. 6d. per ton, and even more where lime is not used as a precipitant. This is irrespective of depreciation of plant and loss of interest on capital. For this 2s. 6d. per ton, about 16 cwt. of water is extracted from the sludge, and it has this one advantage that it can be removed in carts without slopping on the roads or causing any nuisance.

Most of the English authorities on sewage disposal state that the sludge "from any system of sewage disposal consists of about ninety parts of water and ten parts of solids." The sludge which has been allowed to drain dry in the old-fashioned flat-bottomed tanks might possibly have this composition; but as sludge is removed from precipitation tanks it rarely contains less than 94 per cent. of water; indeed, one would be inclined to say that average sludge contained about 95 per cent. of water—that is to say, only half the solid matter it is generally supposed to contain. Sludge removed from the Candy tank contains about 97 per cent. of water and 3 per cent. of solid matter, while that from the conical-

DISPOSAL OF SLUDGE.

bottomed Dortmund tank contains from 93 to 95 per cent. of water, and from 5 to 7 per cent. of solids.

It should be understood that these percentages refer only to the sludge as removed from the tanks; after it has been standing a little while the water would separate, and the residue would undoubtedly only contain 90 per cent. of water and 10 per cent. of solids; after pressing, the sludge contains from 55 to 60 per cent. of water, while air-dried sludge from Birmingham or other places contains about 12 or 15 per cent. of water.

If, therefore, sewage disposal works have sufficient land—some 200 or 300 yards from dwelling houses —for the sludge to be dried upon, this method of dealing with it is cheaper and more satisfactory than by pressing it; as, after pressing, there is still a mass of solid cake equal to one-fifth the bulk of the original sludge to be got rid of, and, in the author's opinion, it is not as a rule worth paying 2s. 6d. per ton to get this comparatively small benefit. Where, however, for any special reason the sludge is offensive, it will probably be necessary to adopt pressing.

CHAPTER VII.

FILTRATION OR NITRIFICATION.

Intermittent land filtration—Artificial filters suggested by Warington in 1882—Conditions necessary for nitrification—Experiments by the Massachusetts Board of Health with coarse sand, fine sand, peat, silt, garden soil, and fine gravel—Results obtained — London County Council experiments — Results obtained.

In the First Report of the Rivers Pollution Commissioners, in 1870, mention is made of downward intermittent filtration of sewage through land, and a number of experiments on the filtering capacity of different soils were carried out.

As the result of laboratory experiments of Sir Edward Frankland, the following conclusions were arrived at :—

1. That the soil should not be too open, so that any quantity of sewage could pass through.

2. That it should be deeply drained to a depth of six feet.

3. That the sewage should be dealt with intermittently.

In the Second Report of the Rivers Pollution

INTERMITTENT FILTRATION.

Commissioners, intermittent filtration was defined as "concentration of sewage *at short intervals* on an area of specially chosen porous ground, as small as will absorb and cleanse it; not excluding vegetation, but making the produce of secondary importance. The intermittency of application is a *sine quâ non* even in suitably constituted soils."

Frankland claimed that one acre of land laid out as an intermittent filtration area was sufficient for a population of 3000 or 4000, or for from 50,000 to 100,000 gallons of sewage a day, the land being divided into about twelve parts, each part receiving successively the whole of the sewage for about six hours. In practice, it was found that intermittent filtration areas could best be worked by land laid out in ridge and furrow, and by growing crops on the ridges, with open sandy soil, worked as a sewage filter without any regard to cropping, possibly this quantity might be purified after precipitation.

At Merthyr Tydvil and at Kendal, this estimate was tried on a practical scale, and it was found that a population of 1000 to the acre was all that could be efficiently dealt with.

The analyses of the sewage before and after filtration at Merthyr Tydvil are not comparable, because a very large quantity of the subsoil water finds its way into the subsoil drains, and dilutes the effluent, apparently giving a percentage of purification which is quite unwarranted.

At Kendal, some 750,000 gallons of sewage, half of which is subsoil water, are dealt with each day by downward intermittent filtration. In the first instance, upon 5½ acres of filter, and subsequently it was found necessary to acquire another 11½ acres.

These two instances of intermittent filtration were carried out some twenty years ago, and since then, intermittent filtration areas have been laid out on most sewage farms, notably at Nottingham and Birmingham. At the latter place, it has been found that, with a gravelly subsoil, the limit of the population that can be dealt with is about 500 to the acre, even after the solid matters have been removed from the sewage by precipitation.

The next step in the evolution of the present biological sewage filter was due to Warington, who in 1882 communicated a paper to the Society of Arts, and pointed out that dilute solutions of ammonium salts or of urine would not undergo nitrification if they were sterilized by boiling, or by the addition of antiseptics, and were supplied with air which was filtered through cotton wool. If, however, a small particle of fresh soil were added, in a little while active nitrification would set in. He further found that this process went on best in the dark, that an alkaline base such as lime was necessary, and a temperature ranging from 40° to 120° Fahr. He further called attention to the researches of MM. Schloesing and Muntz, who claimed that nitrification proceeded

ARTIFICIAL FILTERS.

with the greatest rapidity at a temperature of 99° Fahr. Warington in his article wrote—

"Though, however, porosity is by no means essential to the nitrifying power of a soil, it is undoubtedly a condition having a very favourable influence on the rapidity of the process; a porous soil of open texture will present an immense surface, covered with oxidizing organisms, and generally well supplied with the air requisite for the discharge of their functions. . . . A filtering medium of pure sand and limestone, treated intermittently with sewage, will, after a time, display considerable nitrifying powers, the surfaces becoming covered with oxidizing organisms derived from the sewage. . . . It will be gathered from the observations now made, that it would be possible to construct a filter bed, having a greater oxidizing power than would be possessed by an ordinary soil and subsoil. Such a bed would be made by laying over a system of drain pipes a few feet of soil, obtained from the surface (first six inches) of a good field, the soil being selected as one porous and containing a considerable amount of carbonate of calcium." (See Fig. 12.)

This important suggestion of constructing artificial nitrifying beds or filters lay in abeyance for many years, and was first carried into effect for the purification of large quantities of sewage by the International Sewage Purification Company. The only difference between the filter which Warington suggested and that which the Company constructed was that, instead of using sand and limestone chippings, this Company adopted sand and a magnetic oxide of iron prepared by being roasted with carbon in a closed chamber,

—a preparation which they sold under the trade name of Polarite.

These new artificial filters were tried for eight hours and then allowed to have a rest of sixteen hours. When the quantity of defœcated or clarified sewage dealt with by the filter is not excessive, that is to say is not more than 250 gallons per effective yard per day of eight hours, with a rest of sixteen hours, without any sewage going on, an effluent is, as a rule, produced which contains considerably less than 0·1 part of organic ammonia per 100,000, absorbing less than one part of oxygen at 80° Fahr. in four hours, and containing one part of nitrogen as nitrates. Unfortunately, many of the filters constructed by this Company were constructed to deal with 600, 700, and even 1000 gallons of sewage per square yard, and where these artificial filters have been brought into discredit, it is by their being thus overworked.

The action of the nitrifying filter can perhaps be best understood by comparison with an analogous change which takes place in the fermentation of alcohol into vinegar through the action of the *mycoderma aceti*, and it is worth while to fully consider what the vinegar brewer does when he conducts his operations.

He allows a solution of alcohol to drip slowly over birch twigs in a current of air at a temperature of 77° Fahr. for a fortnight. During this time an organism known as the *mycoderma aceti*, which is seeded on the twigs, flourishes, and, in its growth, takes up oxygen from the air, and unites this with hydrogen from the alcohol, thus forming acetic acid and water. The chemical change which takes place may be represented as follows—

Alcohol + oxygen from the air = acetic acid + water
C_2H_6O +   O   = C_2H_4O + H_2O

I have said that the changes which take place in a sewage filter are analogous to this fermentation, and I have no doubt saltpetre could be produced from sewage, by treating it as the vinegar brewer does his alcohol, but the expense would be prohibitive.

The changes sewage undergoes are, however, more complex. First of all, albuminous matters, and such substances as the urea of the urine (which is the form in which the great bulk of the nitrogen leaves the human body), get converted, into carbonate of ammonia. This change, known as "bacteriolysis," results from the action of the *bacillus ureæ* and other organisms present in the sewage, and is complete by the time the sewage reaches the precipitation tanks.

The change which it is now desired to effect may be regarded, then, as the oxidation by means of nitrifying bacteria of a solution of carbonate of ammonia into the nitrate of some base, generally lime, and this action is known as nitrification.

This is an operation which has been carried on by the Chinese in the manufacture of saltpetre for thousands of years, and the factors necessary for its success are as follows:

1. A substance which will yield ammonia.

2. The presence of calcareous matter to neutralize the nitric acid when formed by the nitrifying organisms; for this reason (except in towns with hard waters) lime should form a constituent of the precipitant.

3. The presence of air in the pores of the filter.

4. A temperature as near 100° Fahr. as possible.

5. A certain amount of time must be allowed for the organisms to grow. On this account the quantity of the sewage applied should never exceed three gallons per square foot of filter per hour, and the filter should only be worked for eight to ten hours a day.

6. As the air in the pores of the filter is consumed, it must be renewed again. This is either done by applying the sewage intermittently, or, as suggested by Mr. Lowcock, by blowing a small quantity of air into the lower part of the filter continuously.

7. The success of the sewage filter will to a great extent depend upon the size of the particles of the sewage filter and its consequent porosity, and to some extent upon the mechanical texture of the material being sufficiently rough to retain the colonies of nitrifying organisms.

THEORY OF NITRIFICATION. 113

When there is a sufficient area of sandy gravel, all these conditions are supplied by nature, except the systematic renewal of the air. This can be arranged by adopting an automatic flushing tank, which will apply the sewage intermittently to the land, the air being drawn into its pores between the flushings; where there is not sufficient fall for one of these contrivances to be adopted, great care must be exercised in frequently changing the areas irrigated.

It will be seen that these filters act chemically and biologically; the dissolved organic matter does not pass out of the filter in the same condition that it enters it—namely, as albuminous matter—but as nitrates. With these coarse-grained artificial filters much more sewage can be passed through in the twenty-four hours than could be passed through land, but little purification would be effected if more than 200 or 300 gallons were filtered. This will be perfectly obvious when we bear in mind that, as has been pointed out, about half the weight of the nitrates produced consists of oxygen, and that therefore, to produce the nitrates, the nitrifying organisms must be freely supplied with oxygen by allowing the filter to empty, and thus cause the air to penetrate into all its interstices.

The necessity for adequately aërating filters was first scientifically worked out by the Massachusetts State Board of Health, who had an experimental station at Lawrence from 1889 to 1893.

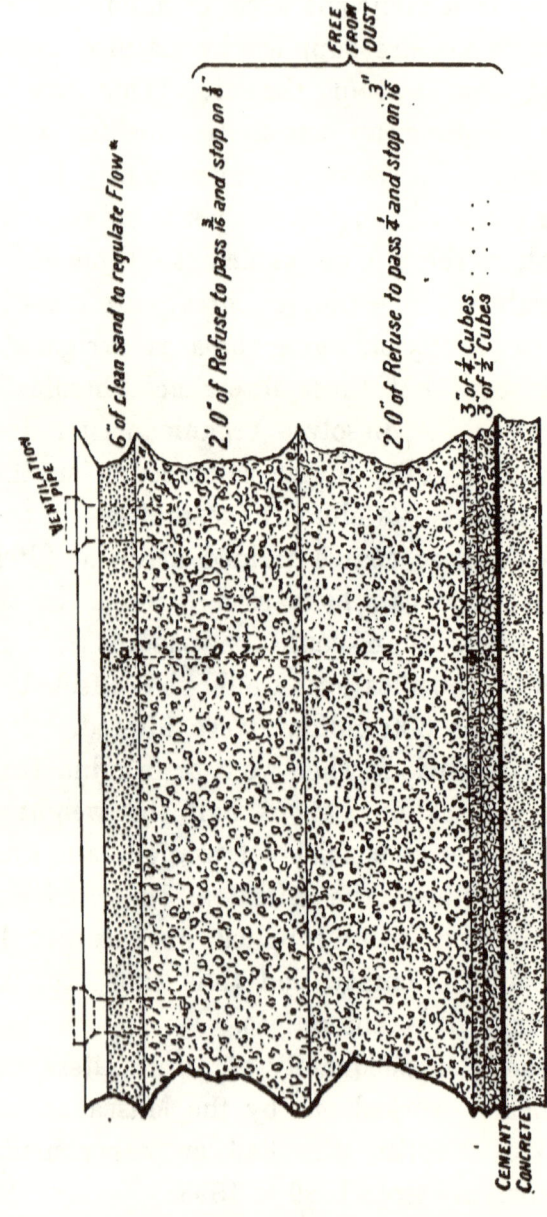

Fig. 12.—Diagram showing Method of constructing Artificial Sewage Filter.

* The rate of Filtration should be such that the water falls in the Filter 6 to 8 inches in one Hour.

The experiments of the Board were made in a series of circular tanks—seventeen feet in diameter and six feet deep. The bottoms of the tanks were drained with ordinary horse-shoe land drains, and large gravel from one inch to two inches in size, covered over with a layer of pea gravel about one-eighth of an inch in size.

The more important experimental tanks were filled as follows—

TANK No. 1.

Five feet of clean coarse sand.

TANK No. 2.

Five feet of fine sand.

TANK No. 3.

Four feet of peat covered with one foot of the original top layer of peat.

TANK No. 4.

Five feet of river silt, very fine sand.

TANK No. 5.

Five feet of garden soil.

TANK No. 6.

Three feet eight inches of coarse and fine sand and fine gravel.

They also experimented with a number of other filters, such as three feet eight inches of coarse sand and fine gravel (same as No. 6), covered with ten inches of sandy loam, and six inches of soil; and three feet eight inches of coarse sand and fine gravel (same as No. 6), covered with eight inches of sandy loam, and eight inches of sandy gravel.

The area of the filters was, roughly speaking, $\frac{1}{200}$

of an acre, and they were used for six days in the week.

The following are the most instructive of the results obtained—

No. 1. **Coarse Sand.**—This filter had a total capacity of 12,300 gallons, and, when filled with water, was found to take up 3240 gallons. When drained away, however, only 2200 gallons of water could be run out, 1040 gallons remaining in the filter owing to capillary attraction; so that when the filter was drained dry it was charged with 2200 gallons of air.

By these experiments the importance of capillarity was first thoroughly brought home; and in the experiments with the fine river silt and garden soil, no nitrification occurred even when the quantity of sewage filtered was only about 7000 gallons to the acre per day. When we recognize the fact that the whole of the nitrification depends entirely upon getting the air in the interstices of the filter changed, the importance of constructing a filter of material of such a size that capillary attraction shall not prevent it from rapidly draining dry will be appreciated.

In filter No. 1, about 100,000 gallons of sewage to the acre were filtered in 24 hours, about 86 per cent. of the albuminoid ammonia being removed, and 89 per cent. of the bacteria.

No. 2. **Fine Sand.**—This filter was only able to deal with 40,000 gallons of sewage to the acre; but although the quantity of sewage that could be purified was less than half that in filter No. 1, it effected a higher degree of purification. The reduction of the albuminoid ammonia was as much as 97 per cent., and as many as 99·8 per cent. of the bacteria were removed.

No. 3. **Peat.**—Without going into detail with regard to tank No. 3, the peat tank, it is sufficient to say that, from the results obtained, the Massachusetts experimenters came to the conclusion that this material was "entirely worthless for the filtration of sewage."

No. 4. **River Silt.**—The results obtained by this filter are similar to those obtained by filter No. 2, the fine sand filter; only, as might be expected, as the material was of finer texture, the quantity of sewage that could be filtered was less. On the other hand, the purification effected was, in some respects, greater. Only 30,000 gallons of sewage per acre could be filtered in 24 hours, and 96 per cent. of albuminoid ammonia was removed, or a little less than in filter No. 2. The smaller chemical purification effected is probably due to the fact that there would be a great difficulty in getting the oxygen into the interstices of the finer filter; but, on the other hand, in this filter, as many as 99·9 per cent. of the bacteria in the sewage were removed.

No. 5. **Garden Soil.**—It was found that the quantity of sewage that could be purified with the garden-soil filter was still less, and the reports state that with a depth of five feet no purification by nitrification took place. Although it was probable that no bacteria came through, the organic matter in the effluent was, at the end of two years, nearly as great as in the sewage. This soil remained continually so nearly saturated that, when only 5000 gallons per acre were being filtered daily, although free to drain over every square foot of the bottom, sufficient air could not be taken in to produce nitrification.

The following Tables show the changes which were effected by fine-gravel filters worked at the rate of 100,000 gallons per acre per day, the sewage being applied fourteen times a day for six days in the week. The air in the interstices of these filters was estimated at one-third their total capacity.

FINE GRAVEL FILTERS.—PARTS PER 100,000.

1889.		Ammonia.			Chlorine.	Nitrogen as	
		Free.	Albuminoid.	Sum of.		Nitrates.	Nitrites.
May 23 to June 22.	Sewage .	1·9919	0·6031	2·5950	5·16	0·0	0·0
	Effluent.	0·0031	0·0375	0·0406	6·00	2·0700	0·0002
	Reduction in albuminoid ammonia 94 per cent.						
June 23 to July 22.	Sewage .	2·2500	0·7255	2·9755	7·46	0·0	0·0
	Effluent.	0·0050	0·0354	0·0404	9·0104	2·2500	0·0004
	Reduction in albuminoid ammonia 95 per cent.						
Sept. 24 to Oct. 24.	Sewage .	2·0559	0·6453	2·7012	5·55	0·0	0·0
	Effluent.	0·0068	0·0325	0·0393	6·42	0·5700	0·0003
	Reduction in albuminoid ammonia 95 per cent.						

The number of bacteria per cubic centimetre was reduced in the following proportion :—

June 23 to July 22.	In the sewage ... In the effluent ... Reduction per cent.	1,813,500 13,523 99·3
Sept. 24 to Oct. 24.	In the sewage ... In the effluent ... Reduction per cent.	3,034,000 11,592 99·6

The American experiments, however valuable, were only made with weak sewage, and upon a very small scale. They demonstrated clearly the necessity for making the supply of sewage intermittent, and showed a percentage of purification which had never been obtained before.

The London County Council, however, repeated the experiments in a modified way on a much larger scale. They constructed one filter of coke breeze (or rather what is technically known as "pan waste") an acre in extent. They also tried smaller filters of the following materials :—

1. Burnt ballast.
2. Lowestoft shingle, pea size.
3. Coke breeze.
4. Sand.
5. A filter containing a certain amount of polarite.

The London sewage was first of all precipitated in the ordinary way with lime water, and the tank effluent thus obtained was filtered.

The average results from the smaller filters are given.* below:—

Description of filter.	Average oxygen absorbed in four hours. Grains per gallon.		Average albuminoid ammonia. Grains per gallon.		Average purification effected, as determined by oxygen absorbed.
	Crude effluent.	Filtrate.	Crude effluent.	Filtrate.	
					Per cent.
Burnt ballast	1·881	1·072	0·243	0·125	43·3
Pea ballast	1·881	0·880	0·257	0·142	52·3
Coke breeze	1·881	0·711	0·262	0·103	62·2
Sand	1·725	1·001	0·250	0·132	46·6
Proprietary filter	1·881	0·721	0·267	0·106	61·6

In the face of these results, a large filter, an acre in extent, was put down. This large filter was composed of three feet of coke breeze and three inches of gravel. After a number of preliminary experiments, the quantity of sewage filtered was gradually increased until as much as 1,000,000 gallons of the tank effluent were filtered per acre per day. As a matter of fact, the quantity filtered was 1⅙ million gallons per day for six days, the filter resting from ten on Saturday till six on Monday morning. The filter was worked under the supervision of Mr. Dibdin, F.I.C., the late chemist to the London County Council, and the method of working it, which is original, was as follows:—

The outlet from the filter was closed, and the

* The results obtained by the London County Council have been published as separate reports; they are also succinctly epitomized in "The Purification of Sewage and Water," by W. J. Dibdin, F.I.C.

LONDON EXPERIMENTS.

sewage was allowed to run on to the filter until it became quite full, its surface being submerged; this generally took about two hours. The filter was then allowed to remain standing, without any more sewage going on, for just one hour, as Mr. Dibdin is of the opinion that the nutrifying organisms require a definite time in which to effect nitrification. After it had been standing for one hour, it was allowed to drain dry. This part of the process took about five hours, so that the whole of the process of filtration took eight hours.

The amount of nitrification effected as given by Dibdin is shown in the Table on next page.

In conclusion, it should be borne in mind that the coarse-grained filters which are worked intermittently effect the largest amount of oxidation or chemical purification. On the other hand, the finest grained sand filters, worked continuously, although effecting no chemical purification, arrest the largest quantity of bacteria.

The most perfect system of purification would consist of passing the sewage first of all through an intermittent filter of coarse coal to thoroughly oxidize the ammonia into nitrates, and then to filter the effluent continuously through fine sand or river silt to arrest the bacteria. Such a dual process is indicated for works above the intake of any water company.

SEWAGE PURIFICATION.

RESULTS OF EXPERIMENTAL FILTER OF COKE BREEZE OF THE LONDON COUNTY COUNCIL.[*]

Date.	Gallons per acre per day.	Oxygen absorbed in four hours at 80° Fahr.			Albuminoid ammonia.			Nitrogen as nitrates. Grains per gallon.	
		Grains per gallon.		Percentage of purification.	Grains per gallon.		Percentage of purification.		
		Tank effluent.	Filtrate.		Tank effluent.	Filtrate.		Tank effluent.	Filtrate.
1894. April 7 to June 9 ...	500,000	4·096	0·856	79·3	0·416	0·095	77·2	0·1280	0·2378
August 3 to November 9 ...	600,000	3·608	0·730	79·6	0·396	0·113	71·4	0·0223	0·1414
November 16, 1894, to March 2, 1895	1,000,000	4·113	0·935	77·5	0·382	0·114	70·2	0·3056	0·6990
1895. April 8 to April 20	1,000,000	3·512	0·884	75·4	0·360	0·102	71·7	0·1431	0·7700

[*] "Purification of Sewage and Water," W. J. Dibdin, p. 57.

CHAPTER VIII.

SPECIAL FORMS OF SEWAGE FILTERS.

Ducat's filter—Garfield's coal filter—Sewage distributors—The Lowcock filter—Comparative results with coke, breeze, coal, sand, and Lowcock's filter, Tipton and Buxton—Automatic arrangements for applying sewage intermittently.

Ducat's Filter.—A further improved method of filtering has been brought out by Colonel Ducat, late Inspector of the Local Government Board. The peculiarity of this filter is, that it is capable of filtering crude sewage. Its sides are constructed of ordinary drain pipes so placed that the outer ends are higher than the inner ones. At certain levels layers of drain pipes go through the whole of the filter. The sewage is caused to trickle *continuously* from carriers going over its surface. By this arrangement it is claimed that the air passes continually through the filter laterally, and that in consequence it will not require rest, and need not be worked intermittently.

The author had the opportunity of seeing this filter at work on September 3rd, 1897. The crude sewage going on yielded 0·72 part of albuminoid ammonia in 100,000, while the filter effluent contained 0·09 part of albuminoid ammonia, and as much as 2·5 parts of nitrogen as nitrates. The percentage of purification effected was 87·5. The quantity of sewage filtered per square yard of filter was 250 gallons per day of 24 hours.

It should be borne in mind that Colonel Ducat, at Hendon, is dealing with raw sewage which has not been precipitated; the depth of the filter, however, is eight feet. Whether better results still would be effected by constructing the filter of coal is a point which is worth experimenting upon. Another arrangement which Colonel Ducat has made is for the filter to be kept warm in the winter months by passing hot-air pipes through it, the air being heated by a slow combustion stove. In a paper read at the Glasgow Conference of the Bristol Institute of Public Health, in September, 1895, the author suggested the advisability of keeping all sewage filters warm by the circulation of hot-water pipes at a temperature of 90° Fahr.

Since the London County Council experiments were made, a considerable number of filter experiments have been carried out in Derbyshire and Staffordshire.

Coal Filters (Garfield's Patent).—Mr. Garfield, engineer of the Sewage Disposal Works, Wolverhampton, and Mr. E. W. T. Jones, County Analyst, Staffordshire, have made a large number of experiments with different filtering media, and they found that the most striking results could be obtained by filtering a tank effluent through coal. The reason why coal should give such excellent results is by no means settled, but there can be no doubt as to the high percentage of purification effected by this medium. The size of the material is a factor of the greatest importance. The authors of the coal-filtering process found that the best results are obtained with filters of a depth not greater than five feet. The bottom three inches of the filter is constructed of inch cubes covered with half-inch cubes for another three inches, this latter being covered with two feet of coal (*free from dust*) which will pass through a $\frac{3}{8}$-inch screen, but stop upon a $\frac{1}{4}$-inch screen, and upon this is a layer of two feet of coal washed *free from dust*, of a size that will go through a $\frac{1}{4}$-inch screen, but stop on $\frac{1}{8}$-inch screen. The top layer of all is composed of coal containing dust which will go through a $\frac{1}{8}$-inch screen. The object of this layer is to distribute the sewage evenly all over the surface of the filter (see Fig. 12).

Mr. Garfield, like Colonel Ducat, works his filters on an entirely different plan from Mr. Dibdin. Instead of closing the outlet of the filter so as to fill it, he distributes the sewage over the surface of the filter by means of iron troughs or pipes which constantly overflow, or by means of a distributor. This distributor will work with a head of three inches and is also suitable for a head of many feet.

It consists of small galvanized tubes (say ¾-inch in diameter) resting on the surface of the filter and placed say three feet apart. These tubes are perforated with small holes, the size of which depends on the head of water and quantity to be delivered. These holes are also three feet apart. The whole series of the tubes are connected in the middle of their length by levers to a rod, which can be moved by hand or otherwise, through an angle of 45°, thus the jet is first thrown on one side of the tube and then on the other. The jets on one tube being half way between the jets on the other, a good and even distribution is obtained.

It will be seen that, in the process of filtering sewage, it is a matter of the greatest importance to have the sewage distributed uniformly over the surface of the filters. A glazed sanitary ware sewage distributor has been brought out by Messrs. Wragg of Swadlincote.

A further improvement is the sewage distributor of Mr. Corbett, the Borough Engineer of Salford. This consists of an arrangement for varying the quantity of sewage from time to time by an automatic flushing syphon, the sewage itself being distributed over the surface of the filter by means of iron pipes three or four feet apart, and having ⅛-inch to ¼-inch holes every three or four feet in their length. Two inches above each hole is an iron plate upon which the sewage impinges, and is thrown in a circle all round.

SEWAGE DISTRIBUTIONS.

As the head in the flushing tank diminishes the radius of the circle narrows, and in this way every part of the filter is alternately sprinkled. See Fig. 13.

Besides the coal filter, which up to the present

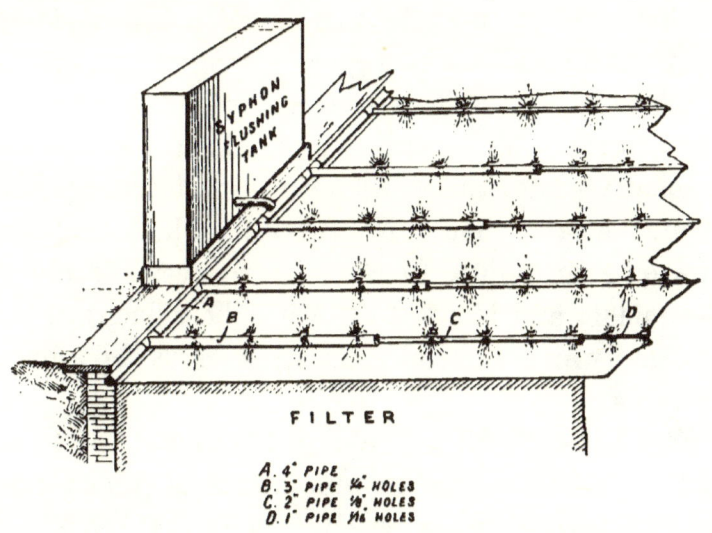

A. 4" PIPE
B. 3" PIPE ¼" HOLES
C. 2" PIPE ⅛" HOLES
D. 1" PIPE 1/16" HOLES

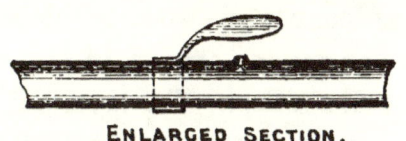

ENLARGED SECTION.

FIG. 13.—AUTOMATIC SEWAGE DISTRIBUTOR.

time has given the best results ever obtained, the Lowcock filter has recently been tried in Staffordshire.

The Lowcock Sewage Filter.—In this most striking innovation in sewage filtering, the principle is adopted of artificially aërating the filters by means of a small blower, and thus increasing their efficiency by supplying the air needed for the sustenance of bacterial action. It is obvious that if this could be accomplished, instead of having filter beds in triplicate, to be used in turn, a filter bed could be used continuously, and a great saving of money and space could thereby be effected. The idea seems to have struck Mr. Lowcock, of Birmingham, and Colonel Waring, the well-known American engineer, about the same time. The two filters, however, have this difference—Mr. Lowcock introduces his air at the middle and blows it out with the effluent at the bottom, while Colonel Waring aërates his filters upwards.

The Lowcock filter (see Fig. 14) consists of three inches of sand resting on nine inches of pea breeze; this in turn rests on twelve inches of bean-sized pebbles, in the middle of which are a number of perforated pipes connected with a blower, which supplies a slow continuous stream of air evenly distributed all over the filter. The layers of the filter above the air are more closely packed than below, so that the sewage and air pass as frothy films downwards over three feet of coke breeze, or other suitable material, to the large open outlet at the bottom.

The pressure of the air necessary to accomplish this is not more than equal to a column of water three or four inches in height. It is obvious that

LOWCOCK'S FILTER.

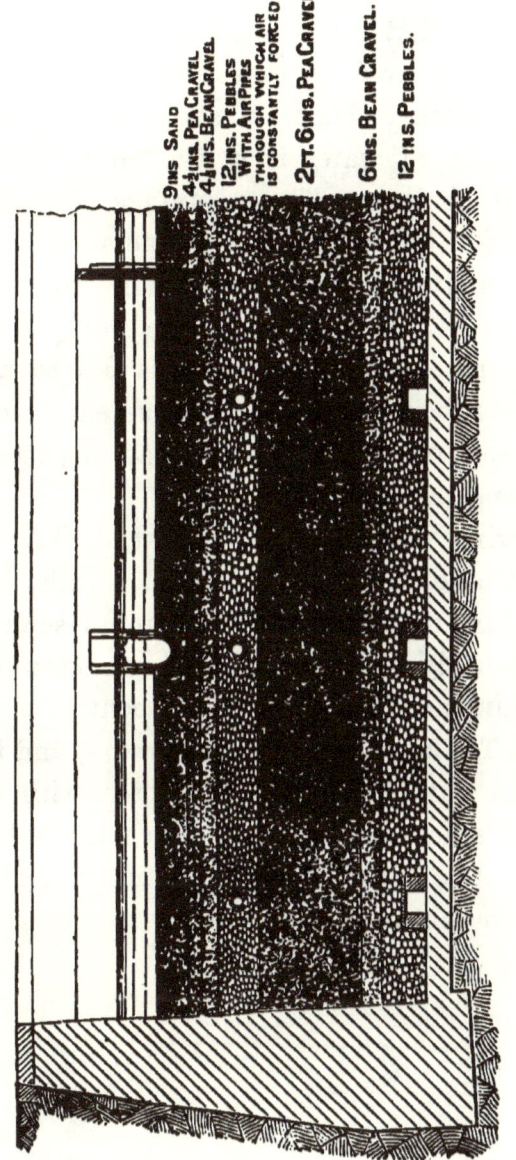

Fig. 14.—Section through Lowcock's Filter.

if there is nine inches of water on the top of the filter the air will not pass up; but I believe the filters are now prepared by selecting the material of gradually increasing size from above downwards, so that, even when no water is in the filter, the air passes out at the bottom. The advantage of this is that it leaves the top of the filter undisturbed to intercept any matters left in suspension from the precipitating tanks and clarifiers. The work of the filter can be increased where there is fall to any depth, more air being supplied at different levels. The result is, that oxidation and nitrification proceed continuously and rapidly.

Colonel Waring was able, at Newport, U.S.A., to effect a purification of 92·5 per cent.; while Mr. Lowcock, dealing with the Wolverhampton sewage, which is extremely difficult to treat, reduces the albuminoid ammonia from 80 to 90 per cent.

Comparative Trials.—The coal, coke breeze, and the Lowcock filters have been tried at Tipton with the following results, for the report of which the author is indebted to Dr. George Reid, the County Medical Officer of Staffordshire :—

RESULTS OF COMPARATIVE TESTS. 131

Sample.	Number of samples analyzed.	Parts per 100,000.								Percentage purification.	
		Solids.			Chlorine.	Free ammonia.	Organic ammonia.	Oxygen absorbed in 4 hrs. at 80° Fahr.	Nitric nitrogen.	On organic ammonia basis.	On oxygen absorbed basis.
		In solution.	In suspension.	Total.							
Tank effluent	13	82·7	1·6	84·3	10·2	1·25	0·23	0·77	Nil	—	—
Old large sand filter	15	76·6	1·5	78·1	10·8	1·19	0·14	0·53	0·09	38·2	31·2
Experimental { Coke breeze	10	84·0	0·9	84·9	10·0	0·90	0·16	0·58	0·38	33·2	29·0
Lowcock's	8	80·7	1·4	82·1	10·0	0·27	0·05	0·22	0·74	75·7	68·5
Garfield's coal	8	91·4	0·3	91·7	10·6	0·19	0·04	0·20	0·81	80·6	70·8

The following are some results obtained with coal, destructor breeze, and coke breeze filters at Buxton:—

	Parts per 100,000.			
	Tank Effluent.		Filtered Effluent.	
	Organic ammonia.	Nitrogen as nitrates.	Organic ammonia.	Nitrogen as nitrates.
Destructor breeze filter	0·23	Nil	0·10	0·10
Coke breeze filter	0·23	Nil	0·11	0·09
Coal filter	0·23	Nil	0·05	0·42

In addition to being tried at Buxton and Tipton, coal filters have been tried at Lichfield, Chesterfield, and Kimberley with equally striking results.

Effluents, such as those from the coal filters at Tipton and Buxton, containing a large quantity of nitrates and little organic ammonia, are chemically practically perfect.

Another improvement which ought to be mentioned with regard to the question of sewage filtration, is the application of the sewage intermittently to the filters by means of automatic flushing tanks. Mr. C. J. Lomax has constructed polarite filters at Failsworth upon this plan.

Instead of having filters in triplicate, using each filter for eight or twelve hours, and then letting it rest for twelve or sixteen hours, each filter is supplied with two automatic flushing syphons, which discharge once every twenty minutes. The tank effluent is allowed to accumulate in a special reservoir for twenty minutes, the syphon then comes into action, and in a minute or two spreads the accumulated clarified sewage over the area of the filter bed, in five minutes the sewage disappears from the surface of the filter and draws air into its interstices. This air remains in contact with the thin films of organic matter left on the surface of the polarite for fifteen minutes, when the syphon again discharges. By placing one's hand at the effluent outlet pipe, just after the syphon has discharged, it is quite easy to feel the air being displaced—in fact, the current is sufficient to blow a match out if the outlet valve is half closed. No analyses are available to show the quantity of oxygen from the air taken up by the sewage, but if the quantity of the sewage is only kept low enough—say, not exceeding 250 gallons per day, per square yard—I believe 80 or 90 per cent. of purification would be effected.

From the number of experiments and actual demonstrations which have been carried out, it has been shown that any degree of purification desired can be effected with sewage filters, by diminishing the quantity of sewage to be filtered per square yard.

A method of making filters work automatically on the Dibdin system is the alternating gear devised by Mr. Cameron, of Exeter. In this arrangement, the effluent from the filter gradually fills a chamber off the effluent drain until the filter is completely waterlogged; the water rising in this chamber actuates a float which releases a valve at the outlet of the filter, and at the same time turns the tank effluent on to another filter. The author has thoroughly investigated this ingenious method of working the filters, and it can be thoroughly relied upon.

More recently still, the simpler method has been adopted of putting an automatic flushing syphon in a chamber in connection with the effluent drain. The effluent from the filter gradually fills this chamber until the filter is waterlogged, it then syphons the filtrate out of the filter very rapidly. This arrangement in practice was found not to work unless ventilating pipes are passed freely through the top part of the filter.

CHAPTER IX.

THE NEW DEPARTURE "BACTERIOLYSIS."

Scott Moncrieff's and Adeney's experiments—The Exeter septic process of Cameron—Results obtained at Exeter—Gases resulting from the septic process.

DURING the last few years attempts have been made by Messrs. Scott Moncrieff and Donald Cameron, of Exeter, to avoid the necessity for precipitation by causing the organic solid matter in suspension in the sewage to become liquefied by means of certain bacteria.

It is well known that many bacteria have the power of liquefying solid albuminous matter; in fact, one of the ordinary methods of distinguishing different species of bacteria is to ascertain their action upon such substances as gelatine, and as many as 196 varieties out of 440 well-known bacteria have this property of liquefying gelatine.

Scott Moncrieff liquefies the organic matter by straining the sewage through shallow channels filled with flints upon the surface of which the liquefying organisms grow.

Mr. Adeney, of Dublin—to whom we owe the word "bacteriolysis"—has suggested that the purification also of sewage could be effected by the action of micro-organisms by adding a sufficient quantity of nitrate of soda to the sewage.

Dr. Sims Woodhead found that in 1 c.c. of Exeter crude sewage there were one million organisms which were anaerobic, or did not grow in the presence of air, and $5\frac{1}{2}$ million organisms which were aerobic, or did live in the presence of air. Of the one million anaerobic, 300,000 were found to be liquefying organisms; and of the $5\frac{1}{2}$ millions aerobic, 500,000 were also found to be liquefying: so that the proportion of liquefying organisms was found to be greater among the anaerobic than among the aerobic.

The rapidity with which the various bacteria liquefy the solid organic matter varies considerably, and it is claimed by Mr. Cameron that those bacteria which live in the absence of air are the most active liquefying organisms.

The Exeter septic process of purification consists of conducting the sewage into large closed tanks capable of holding a day's sewage. The tanks are covered with concrete arches carried by brick piers and levelled over with soil; they are designed to promote the growth of the liquefying organisms present in the sewage, by whose action the organic solids are naturally broken down into simpler substances which can be dealt with by filtration. The flow through the tanks is uniform and continuous, the inlet and outlet being submerged so as to minimize the disturbance of the contents of the tank by the incoming and outgoing streams, and to prevent the admission of air or the exit of gases into the sewers. It is claimed that practically the whole of the organic matter in suspension is by this means rendered liquid.

In the experimental tanks at Exeter, it was found that during the first few months there was an increase in the depth of the deposit in the tank at the rate of an inch per month; but after this, when the sludge has accumulated to a certain extent, the bacteria practically eat it away as fast as it is formed.

In addition to the layer of the sludge at the bottom of the tank, a scum three or four inches in thickness also forms at the top, and flakes of organic matter fall from the surface scum to the bottom of the tank. At the bottom of the tank, therefore, decomposition takes place, and bubbles of gas are carried from this fermenting mass out at the top of the tank, and, in this way, millions of bacteria are continually falling with the organic matter from the scum to the bottom of the tank. Bubbles of gas keep ascending to the top, and carry with them other colonies of bacteria. The whole of the mass in the tank is thus constantly interchanging, and the liquefying bacteria are thus brought in contact with the whole of the organic matter in suspension.

THE EXETER PROCESS.

Of course it is quite impossible for the bacteria to dissolve the road detritus, and reference to Table I. shows that, while the mineral matter in the American sewage was about 5 parts per 100,000, and in the Exeter sewage 10 parts, that in a mining district, like Alfreton—where the quantity of water used is less than 10 gallons per head per day, and there are no water-closets—the mineral matter in suspension is 32·5 parts per 100,000. This suspended matter will have to be got rid of, either by simple subsidence before the sewage enters the septic tank, or else be precipitated in an ordinary Dortmund tank. At Exeter, it is proposed to let this mineral matter subside in three grit chambers, each fifteen feet square, and fifteen feet deep.

After the sewage has passed through the septic tank, it is subsequently filtered through biological nitrifying filters, just as the tank effluent from a precipitating tank is filtered.

At first sight it would appear that, by rendering the solid organic matter in suspension soluble, the amount of organic matter to be nitrified would be increased; and because the sewage would be rendered stronger the difficulty of the problem would be increased.

At Exeter, however, there is no doubt that this is not so; indeed, the total organic matter is actually reduced by the process which takes place in the septic tank.

Mr. Perkins, the Public Analyst of Exeter, found that the average effluent from the septic tank collected throughout the 24 hours yielded 0·66 part of albuminoid ammonia per 100,000, while the raw sewage contained 1·2 parts, the purification effected being nearly 50 per cent.

Dr. Rideal found the albuminoid ammonia was reduced from 1·4 to 0·64 part per 100,000, or a purification of 55 per cent.

The author collected two samples at the same time in dry weather, when the crude sewage was found to contain 0·48 part of albuminoid ammonia per 100,000, and the septic tank effluent contained 0·25 part.

The chlorine at the same time showed that the raw sewage and effluent were similar, while the albuminoid ammonia showed a reduction of nearly 50 per cent.

Messrs. Dibdin and Thudichum found that a purification of 17·5 per cent. was effected in the septic tank.

The actual figures obtained by Mr. Dibdin were the average of twelve samples, the crude sewage yielding 0·212 part of albuminoid ammonia per 100,000, and the tank effluent 0·175 part. The crude sewage is of an exceptionally weak character.

Dr. Dupre found the reduction effected in the organic nitrogen as 27·7 per cent.

The question immediately arises, What becomes

THE EXETER PROCESS.

of the organic matter in the septic tank? The answer to this is, that it leaves the tank in the form of various gases.

Dr. Rideal gives the following two analyses of gases formed in the septic tank :—

	Per cent. by volume.	
Carbonic acid gas, CO_2	0·3	0·6
Marsh gas (methane), CH_4	20·3	24·4
Hydrogen, H	18·2	36·4
Nitrogen, N	61·2	38·6
	100·0	100·0

The gas which is formed is inflammable, and can be burnt through a Bunsen burner; or, in a scheme carried out on a large scale, it could be driven through a water-trap, so as to prevent any risk of explosion, and be burnt under steam boilers, or the heat might be utilized to keep the nitrifying filters warm. In addition to the gases which are driven off from the tank, a considerable amount of carbonic acid passes away dissolved in the tank effluent.

The Table on next page gives the analyses of the raw sewage and tank effluent collected by the writer at 4.45 p.m., on January 3rd, 1898.

It will be noticed that the tank effluent contained a small trace of nitrites. Dr. Rideal also found nitrates in the tank effluent, while Dr. Dupre found that the tank effluent contained from 0·05 to 0·06 part per 100,000 of nitrites as nitric and nitrous acid. This fact is of great importance, because the oxygen in the nitrites certainly did not enter the sewage in the tank, as it is closed as perfectly as it practically can be. It therefore shows that there is in the Exeter sewage a supply of oxygen sufficient to commence the process of nitrification. This condition, which is peculiar and most exceptional in the sewage, is one that is particularly favourable to this process; and wherever similar circumstances exist, the Exeter septic process may be considered as a suitable substitute for precipitation, and the annual cost of chemicals, and the difficulty of dealing with sludge be thereby avoided.

ANALYSES OF RAW SEWAGE AND TANK EFFLUENT
(January 3, 1898).

	Parts per 100,000.					
	Total solids.	Free ammonia.	Albuminoid ammonia.	Nitrites.	Nitrogen as nitrates and nitrites.	Chlorine.
Sewage, 4.45 p.m.	63·0	1·60	0·48	Nil	Nil	5·6
Septic tank effluent, 4.40 p.m.	65·6	4·00	0·25	trace	0·13	6·3
Filtrate, filters 1 and 3, three hours after discharge	65·2	0·185	0·09	Nil	2·00	5·6
Filtrate, filter 4, five minutes after discharge	63·5	0·20	0·09	Nil	2·00	5·5

CHAPTER X.

CONCLUSION.

What process of purification to adopt—When a sewage farm—When precipitation and land—When precipitation and artificial filtration—Indications for a septic process.

THE question with which every Sanitary Authority is brought face to face is the very concrete one of "What process shall we adopt to purify our sewage?"

Before this question can be answered, certain definite knowledge must be obtained as to the nature of the sewage to be treated. Under this heading, information is required as to the nature of the manufacturing wastes turned into the sewage; whether there is a public water supply; the quantity and quality of the water of the district; the quantity of sewage per head per day. If the water supply is obtained from surface wells, the quantity of sewage will probably not be more than ten gallons per head per day. With regard to the water, it should be ascertained whether it is hard or soft, and whether it contains a large quantity of sulphates.

Another important point upon which information must be obtained is, whether the sewers of the district have a good fall so that the sewage will reach the works comparatively fresh, or whether the sewers are flat and the sewage will be in an advanced state of decomposition by the time it reaches the sewage disposal works.

Then another material point will be to ascertain whether much surface water enters the sewers, and an average day's sample of the sewage should be collected for experimental treatment.

Nothing more than broad generalizations can be laid down without definite instances to deal with.

In the first place, let us suppose that we have ordinary domestic sewage to purify, one not containing any manufacturers' wastes. The first investigation that should be made is as to whether there is a sufficient area of good sandy or gravelly soil which can be commanded by gravitation. In carrying out this investigation, great assistance will be obtained by carefully studying the geological maps of the district, particularly if published on a scale of six inches to the mile.

Speaking generally, if the locality is on the boulder or other clay, the new red marl, the coal measures, Yoredale and limestone shales, the mountain limestone, and the pre-Cambrian rocks, the land may be pronounced as not suitable for sewage farm purposes. If, on the other hand, the formation is blown sand, drift, or other gravel, flint and chalk, bunter sandstone, old red sandstone, or possibly the millstone grit, a sewage farm may be advocated.

The next step will be to have trial holes dug to a depth of six feet; too many of the holes cannot be dug, as it is frequently found that patches of sand or gravel, particularly if by the side of a river, do not extend over such large areas as they apparently seem to.

What will be found very useful in carrying out these investigations is a small boring tool five or six feet long, and half-an-inch to three-quarters of an inch in diameter. With this instrument, the nature of the soil and subsoil may be ascertained, without obtaining leave to dig trial holes; and in the course of a couple of hours, with the assistance of a strong man, ten or more trial holes may be made.

Whatever process of purification is adopted, road detritus must be allowed to settle in detritus tanks, and gross suspended matters must be separated by screening the sewage through a half-inch screen. The best screening arrangement, and the simplest, is the cage screen in use at Buxton, illustrated in Fig. 7.

Having come to the conclusion that a sufficient area of sandy or gravelly soil is available, and that the land can be obtained for not more than £150 per acre, a sewage farm may be decided upon. The question will then arise as to whether it will be necessary to precipitate. The answer to this question will depend very largely upon the freshness and nature of the sewage. If it contains a lot of suspended matter, and is quite fresh, the sewers having good gradients, it is desirable that at least a rough kind of precipitating tank should be constructed. The tank effluent need not be of that purity that it is necessary to attain where biological filters alone are contemplated. If the sewage is stale by the time it reaches the disposal works, instead of a precipitating tank, a large settling tank and a rough strainer of coke breeze may be used. In any case it will always be found desirable to have laid out on the sewage farm a small biological intermittent filter, so that when the crops do not require the sewage it can be turned upon the biological filters, and the other arrangements described in Chapter V. should be carried out.

Where the land of the district is suitable for the purification of the sewage, but the price is more than £150 per acre, it will be found desirable to take a smaller area and adopt precipitation.

If the sewage is domestic and weak, either from there being a very large water supply, or because a large quantity of subsoil water finds its way into the sewers, and the water of the district is soft and does not contain many sulphates, such a method as the septic tank process may be indicated.

Where, however, there is not suitable land, or the sewage is not exceptionally weak, the process will have to be one of precipitation and filtration through biological filters.

CONCLUSION.

With regard to precipitation, the best results will be obtained in the Dortmund tank. As to the chemicals which should be used, this will depend entirely upon the chemical nature of the sewage to be treated, and is a matter for definite experimentation in the actual tanks themselves after the construction of the outfall works.

When, however, the sewage contains manufacturing wastes, in almost every case precipitation will have to be adopted—the particular form of precipitant to be used varying with the nature of the manufacturing processes carried on in the district. As has been pointed out in the chapter on Precipitation, in some few cases, such as where there is a large proportion of brewery waste or tannery waste in the sewage, it will be necessary for the sewage to undergo special modifications of the process of purification.

First of all, dealing with manufacturers' wastes, all solid matters in suspension should be kept out of the sewers, and the wastes should be admitted to the sewers *continuously* throughout each working day. Under various circumstances it may be necessary to pass the sewage through a septic tank, and thus remove a large proportion of the organic matter by fermentation; and under some circumstances it must be remembered that the effluent from the septic tank will practically be a solution of sulphuretted hydrogen in sewage, and it will be necessary to subsequently precipitate with lime, or copperas, or a mixture of both, so as to prevent nuisance. After this, the tank effluent will have to be intermittently filtered through biological filters.

Again, under special circumstances it may be necessary to precipitate first, and then pass the sewage through a septic tank; and, in the third place, through biological filters. But these are only special cases, where a considerable amount of manufacturers' wastes, containing a quantity of organic matter, are admitted into the sewers.

Finally, for works above the intakes of a water company, after the sewage has been chemically purified by intermittent filtration, it is desirable that it should be bacteriologically purified by filtration through shallow filters of sand or fine silt.

In any case, before the engineer prepares his plans of construction, the scientific principles upon which the sewage is to be purified should be settled in consultation with a chemist and bacteriologist who has had special experience in dealing with this question. When our Local Authorities deal with this question in the manner advocated, we shall hear of fewer instances of failure and consequent disappointment.

INDEX.

A B. C. process, 91, 102
Acid, influence of, on nitrification, 108
Aëration of filters, 23
Alcohol, fermentation of, to vinegar, like nitrification, 111
Albuminoid ammonia, meaning of, 15
Alum as a precipitant, 88, 102
Alumino-ferric, 91, 102
Amines process, 91
Analyses of average sewage, 12, 13
——— ——— ———, after precipitation and filtration, 24
——— good effluent, 22
——— ——— farm effluents, 27
———, sewages from Lawrence, U.S.A., Buxton, Exeter, London, twenty midden towns, thirty-six w.-c. towns, Salford, Chesterfield, Derby, Alfreton, a " slop-closet " village, Wolverhampton, Burton-on-Trent, Berlin, 29
——— night and day sewage, Bury, 31

Analyses—*continued*
———, deposit from Manchester Ship Canal, 42
——— sludge, 104
——— effluents from fine gravel filters (Massachusetts), 118
——— ——— ——— coke breeze, burnt ballast and sand filters, 120
——— ——— ——— coke breeze filters (L.C.C.), 122
——— ——— ——— coal, sand, coke breeze and Lowcock's filters, 131
——— ——— ——— Exeter process, 140
Analysis, explanation of terms used in, 14–19
Antiseptics, harmful in sewage, 92, 108
Automatic aërators for filters, 127, 132, 133

B ACILLITE process, 92
Bacillus anthracis, vitality of, in water, 44

Bacillus coli communis in sewage, 48
———, *typhosis*, vitality of in water, 44
———, *tuberculosis*, vitality of in water, 44
Bacteria in river waters, 45
——— sewage, 44, 135
———, aerobic and anaerobic, 135
———, liquefying organic matter, 134
———, removal of, by fine sand filters, 118
———, ———, by coarse nitrifying filters, 119
Bacteriolysis, 135
Berlin sewage farm, 76
Birmingham, disposal of sludge at, 103
Buxton, screening arrangements at, 65
———, experimental filters at, 131

CALCULATION of sewage flow from chlorine, 19
——— of maximum discharge of sewers, 40
Cameron, Mr. Donald, 134
Candy's tank, 100
Catchwater system, 63
Chemicals, chiefly used as precipitants, 87–92
Chlorine, significance of, 16, 17
——— in sewage, 18
———, hourly variations in, 32
Cholera, vitality of bacillus of, in water, 44
Clarification of sewage, 64

Clay lands, irrigation of, 56
———, burnt, filters of, 120
Cooper, Mr. C. H., 58
Composition of sewage, 12, 13, 24
Conical-bottomed tanks, 98
Continuous - flow precipitation tanks, 95
Copperas as a precipitant, 88, 90
Cost, relative, of precipitants, 88
——— of sludge pressing, 104
Craigentinny meadows, 73
Crops for sewage farms, 71

DIBDIN on precipitants, 87
Dibdin's method of working filters, 121
Disconnection of house drains from sewer, 8
Drain, definition of, 3

EDINBURGH sewage farm, 73
Effluents: see "Analyses"
Electrolysis of sewage, 92
Exeter process, 135

FAIRLEY, Mr. W., 73
Farms, sewage, 53
——— ———, proper manner of laying out, 70
———, ———, Berlin, 29, 76
———, ———, Birmingham, 71
———, ———, Burton-on-Trent, 29, 67, 93
———, ———, Edinburgh, 73

INDEX.

Farms, sewage, Leicester, 63
——, ——, Nottingham, 73
——, ——, Paris, 76
——, ——, Stratford-on-Avon, 68, 93
Ferrozone, 92
Filters, Massachusetts experiments with, 115, 119
——, coarse sand and fine sand, 116
——, peat, river silt and garden soil, 117
——, fine gravel, 118
——, burnt ballast, coke breeze, sand and polarite (London County Council experiments), 120
——, results with coke breeze, 122
——, Ducat's, 123
——, coal, 124
——, Lowcock's, 127
——, comparative tests of, in Derbyshire and Staffordshire, 131
Filtration, rate of, 112
——, theory of, 109-112
——, bacterial, 118
Fish, destruction of, by sewage, 46
Flow, maximum flow of circular sewers, 40
——, hourly, of sewage, 34

GARFIELD'S filter (coal), 124-131
Gases resulting from decomposition of sewage, 37, 139

HANSON'S process, 92
Hermite process, 92
Herring brine process: *see* Amines process, 91
House drains, disconnection of, 8

INTERNATIONAL Purification Co.'s process, 109
Iron salts as precipitants, 86-88
Irrigation, 62
——, when to be adopted, 143
——, when to be combined with precipitation, 144
Ives' tanks, 101

KINEBÜHLER'S conical-bottomed tank, 95

LAND treatment of sewage, 53
Life in river waters, 46
Lime as a precipitant, 87-89
Lowcock's filter, 128

MANUFACTURERS' wastes, 90
—— ——, chlorine in, 18
Manurial value of sewage, 81
—— —— ——, fallacy of, 83
Massachusetts experiments on value of precipitants, 88
—— ——, on filtration, 115-119

NATIVE guano process: *see* A. B. C. process, 91

Naylor, Mr. W., 42
Nitrification, theory of, 23, 109-113

ORGANIC ammonia, 15
Oxygen absorbed, 19

PARIS farm, 76
Precipitants, 87-92
Precipitation tanks, absolute rest, 94
——— ———, continuous flow, 95
Pressing of sludge, cost of, 104
Ptomaine poisoning through sewage, 48

RIDEAL, Dr., 139
Ridge and furrows, 61
River pollution, 41
———, effect of various degrees of, 41-52
Roechling, C.E., Mr., on Berlin farm, 77

SAND filters, 131
Scott Moncrieff, Mr., 134
Sewage, composition of average, 12

Sewage—*continued*
———, composition of, from various towns, 29
———, bacteria in, 44
———, fluctuating flow of, 33
———, acreage of land for, 80
——— ———, after precipitation, 54-64
Self-purification of rivers, 42-46
Sewage farms: *see* Farms
Sewer, definition of, 3
Sewers, maximum discharge of, 40
Sludge, disposal of, 104
Suspended matter in sewage, 29
Standards of purity for sewage effluents, 20

TANKS, precipitation, 94-101
Tees Valley, outbreak of typhoid, 51
Typhoid fever, due to sewage-contaminated oysters, 49
——— ———, due to sewage contaminated river water, 51

WARINGTON on Nitrification, 109
Water-closets, 6
Waters, solids in various, 11
——— river, bacteria in, 44

THE END.

PRINTED BY WILLIAM CLOWES AND SONS, LIMITED, LONDON AND BECCLES.

Just Published, Crown Octavo, 308 pp. With Numerous Illustrations and Tables. Price 7s. 6d. cloth

WATER AND ITS PURIFICATION

A HANDBOOK FOR THE USE OF LOCAL AUTHORITIES, SANITARY OFFICERS, AND OTHERS INTERESTED IN WATER SUPPLY

BY

SAMUEL RIDEAL, D.Sc.(Lond.)

FELLOW OF THE INSTITUTE OF CHEMISTRY; EXAMINER IN CHEMISTRY TO THE ROYAL COLLEGE OF PHYSICIANS; PUBLIC ANALYST FOR THE LEWISHAM DISTRICT BOARD OF WORKS, ETC.

WITH NUMEROUS ILLUSTRATIONS AND TABLES

"The work is written with singular clearness and freedom from technicalities, the practical side of the question being treated with the same consideration as the theoretical side, including those questions affecting the methods of analysis for determining the fitness of waters for drinking purposes. The last chapter on the Analysis and Interpretation of Results is especially good, summing up the value of both bacteriological and chemical analysis without prejudice for or against the respective merits of these two methods. As a work dealing as clearly as it does with the various ramifications of such an important subject as water and its purification it may be warmly recommended."—*The Lancet*, Feb. 27, 1897.

London: CROSBY LOCKWOOD AND SON
7, STATIONERS' HALL COURT, E.C.

BOOKS FOR SANITARY OFFICIALS.

THE HEALTH OFFICER'S POCKET-BOOK:
A Guide to Sanitary Practice and Law. For Medical Officers of Health, Sanitary Inspectors, Members of Sanitary Authorities, etc. By EDWARD F. WILLOUGHBY, M.D.(Lond.), etc. Fcap. 8vo, 7s. 6d. cloth.

"A mine of condensed information of a pertinent and useful kind on the various subjects of which it treats. The matter seems to have been carefully compiled and arranged for facility of reference, and it is well illustrated by diagrams and wood-cuts. The different subjects are succinctly but fully and scientifically dealt with."—*The Lancet.*

"Ought to be welcome to those for whose use it is designed, since it practically boils down a reference library into a pocket volume. . . . It combines, with an uncommon degree of efficiency, the qualities of accuracy, conciseness and comprehensiveness."—*Scotsman.*

SANITARY ARRANGEMENT OF DWELLING-HOUSES:
A Handbook for Householders and Owners of Houses. By A. J. WALLIS TAYLER, A.M.Inst.C.E. With Illustrations. Crown 8vo, 2s. 6d. cloth.

"This book will be largely read; it will be of considerable service to the public. It is well arranged, easily read, and for the most part devoid of technical terms."—*Lancet.*

VENTILATION :
A Text-book to the Practice of the Art of Ventilating Buildings. By W. P. BUCHAN, R.P. Crown 8vo, 3s. 6d. cloth.

"Contains a great amount of useful practical information as thoroughly interesting as it is technically reliable."—*British Architect.*

PLUMBING :
A Text-book to the Practice of the Art or Craft of the Plumber. By W. P. BUCHAN, R.P. Seventh Edition, Enlarged. Crown 8vo, 3s. 6d. cloth.

"A text-book which may be safely put in the hands of every young plumber."—*Builder.*

SANITARY WORK IN THE SMALLER TOWNS AND IN VILLAGES.
Comprising:—I. SOME OF THE MORE COMMON FORMS OF NUISANCES AND THEIR REMEDIES. II. DRAINAGE. III. WATER SUPPLY. By CHARLES SLAGG, Assoc.Inst.C.E. Second Edition, Revised. 3s. cloth.

LONDON: CROSBY LOCKWOOD AND SON
7, STATIONERS' HALL COURT, E.C.

7, Stationers' Hall Court, London, E.C.

CROSBY LOCKWOOD & SON'S
Catalogue of
Scientific, Technical and Industrial Books.

	PAGE		PAGE
MECHANICAL ENGINEERING	1	DECORATIVE ARTS	30
CIVIL ENGINEERING	9	NATURAL SCIENCE	32
MARINE ENGINEERING, &c.	17	CHEMICAL MANUFACTURES	34
MINING & METALLURGY	19	INDUSTRIAL ARTS	36
ELECTRICITY	23	COMMERCE, TABLES, &c.	41
ARCHITECTURE & BUILDING	25	AGRICULTURE & GARDENING	43
SANITATION & WATER SUPPLY	27	AUCTIONEERING, VALUING, &c.	46
CARPENTRY & TIMBER	28	LAW & MISCELLANEOUS	47

MECHANICAL ENGINEERING, &c.

THE MECHANICAL ENGINEER'S POCKET-BOOK.

Comprising Tables, Formulæ, Rules, and Data: A Handy Book of Reference for Daily Use in Engineering Practice. By D. KINNEAR CLARK, M. Inst. C.E., Third Edition, Revised. Small 8vo, 700 pp., bound in flexible Leather Cover, rounded corners **6/0**

SUMMARY OF CONTENTS:—MATHEMATICAL TABLES.—MEASUREMENT OF SURFACES AND SOLIDS.—ENGLISH AND FOREIGN WEIGHTS AND MEASURES.—MONEYS.—SPECIFIC GRAVITY, WEIGHT, AND VOLUME.—MANUFACTURED METALS.—STEEL PIPES.—BOLTS AND NUTS.—SUNDRY ARTICLES IN WROUGHT AND CAST IRON, COPPER, BRASS, LEAD, TIN, ZINC.—STRENGTH OF TIMBER.—STRENGTH OF CAST IRON.—STRENGTH OF WROUGHT IRON.—STRENGTH OF STEEL.—TENSILE STRENGTH OF COPPER, LEAD, &c.—RESISTANCE OF STONES AND OTHER BUILDING MATERIALS.—RIVETED JOINTS IN BOILER PLATES.—BOILER SHELLS.—WIRE ROPES AND HEMP ROPES.—CHAINS AND CHAIN CABLES.—FRAMING.—HARDNESS OF METALS, ALLOYS, AND STONES.—LABOUR OF ANIMALS.—MECHANICAL PRINCIPLES.—GRAVITY AND FALL OF BODIES.—ACCELERATING AND RETARDING FORCES.—MILL GEARING, SHAFTING, &c.—TRANSMISSION OF MOTIVE POWER.—HEAT.—COMBUSTION: FUELS.—WARMING, VENTILATION, COOKING STOVES.—STEAM.—STEAM ENGINES AND BOILERS.—RAILWAYS.—TRAMWAYS.—STEAM SHIPS.—PUMPING STEAM ENGINES AND PUMPS.—COAL GAS, GAS ENGINES, &c.—AIR IN MOTION.—COMPRESSED AIR.—HOT AIR ENGINES.—WATER POWER.—SPEED OF CUTTING TOOLS.—COLOURS.—ELECTRICAL ENGINEERING.

"Mr. Clark manifests what is an innate perception of what is likely to be useful in a pocket-book, and he is really unrivalled in the art of condensation. It is very difficult to hit upon any mechanical engineering subject concerning which this work supplies no information, and the excellent index at the end adds to its utility. In one word, it is an exceedingly handy and efficient tool, possessed of which the engineer will be saved many a wearisome calculation, or yet more wearisome hunt through various text-books and treatises, and, as such, we can heartily recommend it to our readers."—*The Engineer.*

"It would be found difficult to compress more matter within a similar compass, or produce a book of 650 pages which should be more compact or convenient for pocket reference. . . . Will be appreciated by mechanical engineers of all classes."—*Practical Engineer.*

L.   ·   A

MR. HUTTON'S PRACTICAL HANDBOOKS.

THE WORKS' MANAGER'S HANDBOOK.
Comprising Modern Rules, Tables, and Data. For Engineers, Millwrights, and Boiler Makers; Tool Makers, Machinists, and Metal Workers; Iron and Brass Founders, &c. By W. S. HUTTON, Civil and Mechanical Engineer, Author of "The Practical Engineer's Handbook." Fifth Edition, carefully Revised, with Additions. In One handsome Volume, medium 8vo, strongly bound **15/0**

☛ *The Author having compiled Rules and Data for his own use in a great variety of modern engineering work, and having found his notes extremely useful, decided to publish them—revised to date—believing that a practical work, suited to the* DAILY REQUIREMENTS OF MODERN ENGINEERS, *would be favourably received.*

"Of this edition we may repeat the appreciative remarks we made upon the first and third. Since the appearance of the latter very considerable modifications have been made, although the total number of pages remains almost the same. It is a very useful collection of rules, tables, and workshop and drawing office data."—*The Engineer,* May 10, 1895.
"The author treats every subject from the point of view of one who has collected workshop notes for application in workshop practice, rather than from the theoretical or literary aspect. The volume contains a great deal of that kind of information which is gained only by practical experience, and is seldom written in books."—*The Engineer,* June 5, 1885.
"The volume is an exceedingly useful one, brimful with engineer's notes, memoranda, and rules, and well worthy of being on every mechanical engineer's bookshelf."—*Mechanical World.*
"The information is precisely that likely to be required in practice. . . . The work forms a desirable addition to the library not only of the works' manager, but of any one connected with general engineering."—*Mining Journal.*
"Brimful of useful information, stated in a concise form, Mr. Hutton's books have met a pressing want among engineers. The book must prove extremely useful to every practical man possessing a copy."—*Practical Engineer.*

THE PRACTICAL ENGINEER'S HANDBOOK.
Comprising a Treatise on Modern Engines and Boilers, Marine, Locomotive, and Stationary. And containing a large collection of Rules and Practical Data relating to Recent Practice in Designing and Constructing all kinds of Engines, Boilers, and other Engineering work. The whole constituting a comprehensive Key to the Board of Trade and other Examinations for Certificates of Competency in Modern Mechanical Engineering. By WALTER S. HUTTON, Civil and Mechanical Engineer, Author of "The Works' Manager's Handbook for Engineers." &c. With upwards of 370 Illustrations. Fifth Edition, Revised with Additions. Medium 8vo, nearly 500 pp., strongly bound. [*Just Published.* **18/0**

☛ *This Work is designed as a companion to the Author's* "WORKS' MANAGER'S HANDBOOK." *It possesses many new and original features, and contains, like its predecessor, a quantity of matter not originally intended for publication, but collected by the Author for his own use in the construction of a great variety of* MODERN ENGINEERING WORK.

The information is given in a condensed and concise form, and is illustrated by upwards of 370 Woodcuts; and comprises a quantity of tabulated matter of great value to all engaged in designing, constructing, or estimating for ENGINES, BOILERS, *and* OTHER ENGINEERING WORK.

"We have kept it at hand for several weeks, referring to it as occasion arose, and we have not on a single occasion consulted its pages without finding the information of which we were in quest.'—*Athenæum.*
"A thoroughly good practical handbook, which no engineer can go through without learning something that will be of service to him."—*Marine Engineer.*
"An excellent book of reference for engineers, and a valuable text-book for students of engineering."—*Scotsman.*
"This valuable manual embodies the results and experience of the leading authorities on mechanical engineering."—*Building News.*
"The author has collected together a surprising quantity of rules and practical data, and has shown much judgment in the selections he has made. . . . There is no doubt that this book is one of the most useful of its kind published, and will be a very popular compendium."—*Engineer.*
"A mass of information set down in simple language, and in such a form that it can be easily referred to at any time. The matter is uniformly good and well chosen, and is greatly elucidated by the illustrations. The book will find its way on to most engineers' shelves, where it will rank as one of the most useful books of reference."—*Practical Engineer.*
"Full of useful information, and should be found on the office shelf of all practical engineers. —*English Mechanic.*

MR. HUTTON'S PRACTICAL HANDBOOKS—*continued.*

STEAM BOILER CONSTRUCTION.

A Practical Handbook for Engineers, Boiler-Makers, and Steam Users. Containing a large Collection of Rules and Data relating to Recent Practice in the Design, Construction, and Working of all Kinds of Stationary, Locomotive, and Marine Steam-Boilers. By WALTER S. HUTTON, Civil and Mechanical Engineer, Author of "The Works' Manager's Handbook," "The Practical Engineer's Handbook," &c. With upwards of 500 Illustrations. Third Edition, Revised and much Enlarged, medium 8vo, cloth . . **18/0**

☞ THIS WORK *is issued in continuation of the Series of Handbooks written by the Author, viz.:*—"THE WORKS' MANAGER'S HANDBOOK" *and* "THE PRACTICAL ENGINEER'S HANDBOOK," *which are so highly appreciated by engineers for the practical nature of their information; and is consequently written in the same style as those works.*

The Author believes that the concentration, in a convenient form for easy reference, of such a large amount of thoroughly practical information on Steam-Boilers, will be of considerable service to those for whom it is intended, and he trusts the book may be deemed worthy of as favourable a reception as has been accorded to its predecessors.

"One of the best, if not the best, books on boilers that has ever been published. The information is of the right kind, in a simple and accessible form. So far as generation is concerned, this is, undoubtedly, the standard book on steam practice."—*Electrical Review.*

"Every detail, both in boiler design and management, is clearly laid before the reader. The volume shows that boiler construction has been reduced to the condition of one of the most exact sciences; and such a book is of the utmost value to the *fin de siècle* Engineer and Works Manager.'—*Marine Engineer.*

"There has long been room for a modern handbook on steam boilers; there is not that room now, because Mr. Hutton has filled it. It is a thoroughly practical book for those who are occupied in the construction, design, selection, or use of boilers."—*Engineer.*

"The book is of so important and comprehensive a character that it must find its way into the libraries of every one interested in boiler using or boiler manufacture if they wish to be thoroughly informed. We strongly recommend the book for the intrinsic value of its contents."—*Machinery Market.*

PRACTICAL MECHANICS' WORKSHOP COMPANION.

Comprising a great variety of the most useful Rules and Formulæ in Mechanical Science, with numerous Tables of Practical Data and Calculated Results for Facilitating Mechanical Operations. By WILLIAM TEMPLETON, Author of "The Engineer's Practical Assistant," &c., &c. Seventeenth Edition, Revised, Modernised, and considerably Enlarged by WALTER S. HUTTON, C.E., Author of "The Works' Manager's Handbook," "The Practical Engineer's Handbook," &c. Fcap. 8vo, nearly 500 pp., with 8 Plates and upwards of 250 Illustrative Diagrams, strongly bound for workshop or pocket wear and tear . **6/0**

"In its modernised form Hutton's 'Templeton' should have a wide sale, for it contains much valuable information which the mechanic will often find of use, and not a few tables and notes which he might look for in vain in other works. This modernised edition will be appreciated by all who have learned to value the original editions of 'Templeton.'"—*English Mechanic.*

"It has met with great success in the engineering workshop, as we can testify; and there are a great many men who, in a great measure, owe their rise in life to this little book."—*Building News.*

"This familiar text-book—well known to all mechanics and engineers—is of essential service to the every-day requirements of engineers, millwrights, and the various trades connected with engineering and building. The new modernised edition is worth its weight in gold."—*Building News.* (Second Notice.)

"This well-known and largely-used book contains information, brought up to date, of the sort so useful to the foreman and draughtsman. So much fresh information has been introduced as to constitute it practically a new book. It will be largely used in the office and workshop."—*Mechanical World.*

"The publishers wisely entrusted the task of revision of this popular, valuable, and useful book to Mr. Hutton, than whom a more competent man they could not have found."—*Iron.*

ENGINEER'S AND MILLWRIGHT'S ASSISTANT.

A Collection of Useful Tables, Rules, and Data. By WILLIAM TEMPLETON. Seventh Edition, with Additions. 18mo, cloth **2/6**

"Occupies a foremost place among books of this kind. A more suitable present to an apprentice to any of the mechanical trades could not possibly be made."—*Building News.*

"A deservedly popular work. It should be in the 'drawer' of every mechanic.'—*English Mechanic.*

4   CROSBY LOCKWOOD & SON'S CATALOGUE.

THE MECHANICAL ENGINEER'S REFERENCE BOOK.

For Machine and Boiler Construction. In Two Parts. Part I. GENERAL ENGINEERING DATA. Part II. BOILER CONSTRUCTION. With 51 Plates and numerous Illustrations. By NELSON FOLEY, M.I.N.A. Second Edition, Revised throughout and much Enlarged. Folio, half-bound, net . **£3 3s.**

PART I.—MEASURES.—CIRCUMFERENCES AND AREAS, &c., SQUARES, CUBES, FOURTH POWERS.—SQUARE AND CUBE ROOTS.—SURFACE OF TUBES.—RECIPROCALS.—LOGARITHMS. — MENSURATION. — SPECIFIC GRAVITIES AND WEIGHTS.—WORK AND POWER. — HEAT. — COMBUSTION. — EXPANSION AND CONTRACTION. — EXPANSION OF GASES.—STEAM.—STATIC FORCES.—GRAVITATION AND ATTRACTION.—MOTION AND COMPUTATION OF RESULTING FORCES.—ACCUMULATED WORK.—CENTRE AND RADIUS OF GYRATION.—MOMENT OF INERTIA.—CENTRE OF OSCILLATION.—ELECTRICITY.—STRENGTH OF MATERIALS.—ELASTICITY.—TEST SHEETS OF METALS.—FRICTION.—TRANSMISSION OF POWER.—FLOW OF LIQUIDS.—FLOW OF GASES.—AIR PUMPS, SURFACE CONDENSERS, &c.—SPEED OF STEAMSHIPS.—PROPELLERS.—CUTTING TOOLS.—FLANGES. —COPPER SHEETS AND TUBES.—SCREWS, NUTS, BOLT HEADS, &c.—VARIOUS RECIPES AND MISCELLANEOUS MATTER.—WITH DIAGRAMS FOR VALVE-GEAR, BELTING AND ROPES, DISCHARGE AND SUCTION PIPES, SCREW PROPELLERS, AND COPPER PIPES.

PART II.—TREATING OF POWER OF BOILERS.—USEFUL RATIOS.—NOTES ON CONSTRUCTION. — CYLINDRICAL BOILER SHELLS. — CIRCULAR FURNACES. — FLAT PLATES.—STAYS. — GIRDERS.—SCREWS. — HYDRAULIC TESTS. — RIVETING. — BOILER SETTING, CHIMNEYS, AND MOUNTINGS.—FUELS, &c.—EXAMPLES OF BOILERS AND SPEEDS OF STEAMSHIPS.—NOMINAL AND NORMAL HORSE POWER.—WITH DIAGRAMS FOR ALL BOILER CALCULATIONS AND DRAWINGS OF MANY VARIETIES OF BOILERS.

"The book is one which every mechanical engineer may, with advantage to himself, add to his library."—*Industries.*

"Mr. Foley is well fitted to compile such a work. . . . The diagrams are a great feature of the work. . . . Regarding the whole work, it may be very fairly stated that Mr. Foley has produced a volume which will undoubtedly fulfil the desire of the author and become indispensable to all mechanical engineers."—*Marine Engineer.*

"We have carefully examined this work, and pronounce it a most excellent reference book for the use of marine engineers."—*Journal of American Society of Naval Engineers.*

COAL AND SPEED TABLES.

A Pocket Book for Engineers and Steam Users. By NELSON FOLEY, Author of "The Mechanical Engineer's Reference Book." Pocket-size, cloth . **3/6**

"These tables are designed to meet the requirements of every-day use; are of sufficient scope for most practical purposes, and may be commended to engineers and users of steam."—*Iron.*

TEXT-BOOK ON THE STEAM ENGINE.

With a Supplement on GAS ENGINES, and PART II. on HEAT ENGINES. By T. M. GOODEVE, M.A., Barrister-at-Law, Professor of Mechanics at the Royal College of Science, London; Author of "The Principles of Mechanics," "The Elements of Mechanism," &c. Fourteenth Edition. Crown 8vo, cloth . **6/0**

"Professor Goodeve has given us a treatise on the steam engine which will bear comparison with anything written by Huxley or Maxwell, and we can award it no higher praise."—*Engineer.*

"Mr. Goodeve's text-book is a work of which every young engineer should possess himself." —*Mining Journal.*

ON GAS ENGINES.

With Appendix describing a Recent Engine with Tube Igniter. By T. M. GOODEVE, M.A. Crown 8vo, cloth **2/6**

"Like all Mr. Goodeve's writings, the present is no exception in point of general excellence. It is a valuable little volume."—*Mechanical World.*

A TREATISE ON STEAM BOILERS.

Their Strength, Construction, and Economical Working. By R. WILSON, C.E. Fifth Edition. 12mo, cloth **6/0**

"The best treatise that has ever been published on steam boilers.—*Engineer.*

"The author shows himself perfect master of his subject, and we heartily recommend all employing steam power to possess themselves of the work."—*Ryland's Iron Trade Circular.*

THE MECHANICAL ENGINEER'S COMPANION

of Areas, Circumferences, Decimal Equivalents, in inches and feet, millimetres, squares, cubes, roots, &c.; Weights, Measures, and other Data. Also Practical Rules for Modern Engine Proportions. By R. EDWARDS, M.Inst.C.E. Fcap. 8vo, cloth. [*Just Published.* **3/6**

"A very useful little volume. It contains many tables, classified data and memoranda, generally useful to engineers."—*Engineer.*

"This small book is what it professes to be, viz. :—'a handy office companion,' giving as it does, in a succinct form, a variety of information likely to be required by mechanical engineers in their everyday office work."—*Nature.*

A HANDBOOK ON THE STEAM ENGINE.

With especial Reference to Small and Medium-sized Engines. For the Use of Engine Makers, Mechanical Draughtsmen, Engineering Students, and users of Steam Power. By HERMAN HAEDER, C.E. Translated from the German with considerable additions and alterations, by H. H. P. POWLES, A.M.I.C.E., M.I.M.E. Second Edition, Revised. With nearly 1,100 Illustrations. Crown 8vo, cloth **9/0**

"A perfect encyclopædia of the steam engine and its details, and one which must take a permanent place in English drawing-offices and workshops."—*A Foreman Pattern-maker.*

"This is an excellent book, and should be in the hands of all who are interested in the construction and design of medium-sized stationary engines. . . . A careful study of its contents and the arrangement of the sections leads to the conclusion that there is probably no other book like it in this country. The volume aims at showing the results of practical experience, and it certainly may claim a complete achievement of this idea."—*Nature.*

"There can be no question as to its value. We cordially commend it to all concerned in the design and construction of the steam engine."—*Mechanical World.*

BOILER AND FACTORY CHIMNEYS.

Their Draught-Power and Stability. With a chapter on *Lightning Conductors.* By ROBERT WILSON, A.I.C.E., Author of "A Treatise on Steam Boilers," &c. Crown 8vo, cloth **3/6**

"A valuable contribution to the literature of scientific building."—*The Builder.*

BOILER MAKER'S READY RECKONER & ASSISTANT.

With Examples of Practical Geometry and Templating, for the Use of Platers, Smiths, and Riveters. By JOHN COURTNEY, Edited by D. K. CLARK, M.I.C.E. Third Edition, 480 pp., with 140 Illustrations. Fcap. 8vo . **7/0**

"No workman or apprentice should be without this book."—*Iron Trade Circular.*

REFRIGERATING & ICE-MAKING MACHINERY.

A Descriptive Treatise for the Use of Persons Employing Refrigerating and Ice-Making Installations, and others. By A. J. WALLIS-TAYLER, A.-M. Inst. C.E. Second Edition, Revised and Enlarged. With Illustrations. Crown 8vo, cloth. [*Just Published.* **7/6**

"Practical, explicit, and profusely illustrated."—*Glasgow Herald.*

"We recommend the book, which gives the cost of various systems and illustrations showing details of parts of machinery and general arrangements of complete installations."—*Builder.*

"May be recommended as a useful description of the machinery, the processes, and of the facts, figures, and tabulated physics of refrigerating. It is one of the best compilations on the subject."—*Engineer.*

THE LOCOMOTIVE ENGINE AND ITS DEVELOPMENT.

A Popular Treatise on the Gradual Improvements made in Railway Engines between 1803 and 1896. By CLEMENT E. STRETTON, C.E. Fifth Edition, Enlarged. With 120 Illustrations. Crown 8vo, cloth. [*Just Published.* **3/6**

"Students of railway history and all who are interested in the evolution of the modern locomotive will find much to attract and entertain in this volume."—*The Times.*

"The author of this work is well known to the railway world, and no one, probably, has a better knowledge of the history and development of the locomotive. The volume before us should be of value to all connected with the railway system of this country."—*Nature.*

ENGINEERING ESTIMATES, COSTS, AND ACCOUNTS.

A Guide to Commercial Engineering. With numerous examples of Estimates and Costs of Millwright Work, Miscellaneous Productions, Steam Engines and Steam Boilers; and a Section on the Preparation of Costs Accounts. By A GENERAL MANAGER. Second Edition. 8vo, cloth. [*Just Published.* **12/0**

"This is an excellent and very useful book, covering subject-matter in constant requisition in every factory and workshop. . . . The book is invaluable, not only to the young engineer, but also to the estimate department of every works."—*Builder.*

"We accord the work unqualified praise. The information is given in a plain, straightforward manner, and bears throughout evidence of the intimate practical acquaintance of the author with every phase of commercial engineering."—*Mechanical World.*

AERIAL OR WIRE-ROPE TRAMWAYS.

Their Construction and Management. By A. J. WALLIS-TAYLER, A.M. Inst. C.E. With 81 Illustrations. Crown 8vo, cloth. [*Just Published.* **7/6**

"This is in its way an excellent volume. Without going into the minutiæ of the subject, it yet lays before its readers a very good exposition of the various systems of rope transmission in use, and gives as well not a little valuable information about their working, repair, and management. We can safely recommend it as a useful general treatise on the subject."—*The Engineer*.

"Mr. Tayler has treated the subject as concisely as thoroughness would permit. The book will rank with the best on this useful topic, and we recommend it to those whose business is the transporting of minerals and goods."—*Mining Journal*.

MOTOR CARS OR POWER-CARRIAGES FOR COMMON ROADS.

By A. J. WALLIS-TAYLER, Assoc. Memb. Inst. C.E., Author of "Modern Cycles," &c. 212 pp., with 76 Illustrations. Crown 8vo, cloth . . **4/6**

"Mr. Wallis-Tayler's book is a welcome addition to the literature of the subject, as it is the production of an Engineer, and has not been written with a view to assist in the promotion of companies. . . . The book is clearly expressed throughout, and is just the sort of work that an engineer, thinking of turning his attention to motor-carriage work, would do well to read as a preliminary to starting operations."—*Engineering*.

PLATING AND BOILER MAKING.

A Practical Handbook for Workshop Operations. By JOSEPH G. HORNER, A.M.I.M.E. 380 pp. with 338 Illustrations. Crown 8vo, cloth . . **7/6**

"The latest production from the pen of this writer is characterised by that evidence of close acquaintance with workshop methods which will render the book exceedingly acceptable to the practical hand. We have no hesitation in commending the work as a serviceable and practical handbook on a subject which has not hitherto received much attention from those qualified to deal with it in a satisfactory manner."—*Mechanical World*.

PATTERN MAKING.

A Practical Treatise, embracing the Main Types of Engineering Construction, and including Gearing, both Hand and Machine-made, Engine Work, Sheaves and Pulleys, Pipes and Columns, Screws, Machine Parts, Pumps and Cocks, the Moulding of Patterns in Loam and Greensand, &c., together with the methods of estimating the weight of Castings; with an Appendix of Tables for Workshop Reference. By JOSEPH G. HORNER, A.M.I.M.E. Second Edition, Enlarged. With 450 Illustrations. Crown 8vo, cloth **7/6**

"A well-written technical guide, evidently written by a man who understands and has practised what he has written about. . . . We cordially recommend it to engineering students, young journeymen, and others desirous of being initiated into the mysteries of pattern-making."—*Builder*.

"More than 400 illustrations help to explain the text, which is, however, always clear and explicit, thus rendering the work an excellent *vade mecum* for the apprentice who desires to become master of his trade."—*English Mechanic*.

MECHANICAL ENGINEERING TERMS

(Lockwood's Dictionary of). Embracing those current in the Drawing Office, Pattern Shop, Foundry, Fitting, Turning, Smiths', and Boiler Shops, &c., &c. Comprising upwards of 6,000 Definitions. Edited by JOSEPH G. HORNER, A.M.I.M.E. Second Edition, Revised, with Additions. Crown 8vo, cloth **7/6**

"Just the sort of handy dictionary required by the various trades engaged in mechanical engineering. The practical engineering pupil will find the book of great value in his studies, and every foreman engineer and mechanic should have a copy."—*Building News*.

"Not merely a dictionary, but, to a certain extent, also a most valuable guide. It strikes us as a happy idea to combine with a definition of the phrase useful information on the subject of which it treats."—*Machinery Market*.

TOOTHED GEARING.

A Practical Handbook for Offices and Workshops. By JOSEPH HORNER, A.M.I.M.E. With 184 Illustrations. Crown 8vo, cloth . . . **6/0**

"We must give the book our unqualified praise for its thoroughness of treatment, and we can heartily recommend it to all interested as the most practical book on the subject yet written."—*Mechanical World*.

FIRES, FIRE-ENGINES, AND FIRE BRIGADES.

With a History of Fire-Engines, their Construction, Use, and Management; Foreign Fire Systems; Hints on Fire-Brigades, &c. By CHARLES F. T. YOUNG, C.E. 8vo, cloth . . . . . . **£1 4s.**

"To such of our readers as are interested in the subject of fires and fire apparatus we can most heartily commend this book."—*Engineering*.

STONE-WORKING MACHINERY.

A Manual dealing with the Rapid and Economical Conversion of Stone. With Hints on the Arrangement and Management of Stone Works. By M. POWIS BALE, M.I.M.E. Second Edition, enlarged. With Illustrations. Crown 8vo, cloth. [*Just Published.* **9/0**

"The book should be in the hands of every mason or student of stonework."—*Colliery Guardian.*
"A capital handbook for all who manipulate stone for building or ornamental purposes."—*Machinery Market.*

PUMPS AND PUMPING.

A Handbook for Pump Users. Being Notes on Selection, Construction, and Management. By M. POWIS BALE, M.I.M.E. Third Edition, Revised. Crown 8vo, cloth. [*Just Published.* **2/6**

"The matter is set forth as concisely as possible. In fact, condensation rather than diffuseness has been the author's aim throughout; yet he does not seem to have omitted anything likely to be of use."—*Journal of Gas Lighting.*
"Thoroughly practical and simply and clearly written."—*Glasgow Herald.*

MILLING MACHINES AND PROCESSES.

A Practical Treatise on Shaping Metals by Rotary Cutters. Including Information on Making and Grinding the Cutters. By PAUL N. HASLUCK, Author of "Lathe-Work." 352 pp. With upwards of 300 Engravings. Large crown 8vo, cloth **12/6**

"A new departure in engineering literature. . . . We can recommend this work to all interested in milling machines; it is what it professes to be—a practical treatise."—*Engineer.*
"A capital and reliable book which will no doubt be of considerable service both to those who are already acquainted with the process as well as to those who contemplate its adoption.'—*Industries.*

LATHE-WORK.

A Practical Treatise on the Tools, Appliances, and Processes employed in the Art of Turning. By PAUL N. HASLUCK. Fifth Edition. Crown 8vo, cloth **5/0**

"Written by a man who knows not only how work ought to be done, but who also knows how to do it, and how to convey his knowledge to others. To all turners this book would be valuable."—*Engineering.*
"We can safely recommend the work to young engineers. To the amateur it will simply be invaluable. To the student it will convey a great deal of useful information."—*Engineer.*

SCREW-THREADS,

And Methods of Producing Them. With numerous Tables and complete Directions for using Screw-Cutting Lathes. By PAUL N. HASLUCK, Author of "Lathe-Work," &c. With Seventy-four Illustrations. Fourth Edition, Re-written and Enlarged. Waistcoat-pocket size **1/6**

"Full of useful information, hints and practical criticism. Taps, dies, and screwing tools generally are illustrated and their actions described."—*Mechanical World.*
"It is a complete compendium of all the details of the screw-cutting lathe; in fact a *multum-in-parvo* on all the subjects it treats upon."—*Carpenter and Builder.*

TABLES AND MEMORANDA FOR ENGINEERS, MECHANICS, ARCHITECTS, BUILDERS, &c.

Selected and Arranged by FRANCIS SMITH. Sixth Edition, Revised, including ELECTRICAL TABLES, FORMULÆ, and MEMORANDA. Waistcoat-pocket size, limp leather. [*Just Published.* **1/6**

"It would, perhaps, be as difficult to make a small pocket-book selection of notes and formulæ to suit ALL engineers as it would be to make a universal medicine; but Mr. Smith's waistcoat-pocket collection may be looked upon as a successful attempt."—*Engineer.*
"The best example we have ever seen of 270 pages of useful matter packed into the dimensions of a card-case."—*Building News.* "A veritable pocket treasury of knowledge."—*Iron.*

POCKET GLOSSARY OF TECHNICAL TERMS.

English-French, French-English; with Tables suitable for the Architectural, Engineering, Manufacturing, and Nautical Professions. By JOHN JAMES FLETCHER, Engineer and Surveyor. Second Edition, Revised and Enlarged, 200 pp. Waistcoat-pocket size, limp leather **1/6**

"It is a very great advantage for readers and correspondents in France and England to have so large a number of the words relating to engineering and manufacturers collected in a lilliputian volume. The little book will be useful both to students and travellers."—*Architect.*
"The glossary of terms is very complete, and many of the Tables are new and well arranged. We cordially commend the book."—*Mechanical World.*

8   CROSBY LOCKWOOD & SON'S CATALOGUE.

THE ENGINEER'S YEAR BOOK FOR 1899.
Comprising Formulæ, Rules, Tables, Data and Memoranda in Civil, Mechanical, Electrical, Marine and Mine Engineering. By H. R. KEMPE, A.M. Inst. C.E., M.I.E.E., Technical Officer of the Engineer-in-Chief's Office, General Post Office, London, Author of "A Handbook of Electrical Testing," "The Electrical Engineer's Pocket-Book," &c. With about 900 Illustrations, specially Engraved for the work. Crown 8vo, 750 pp., leather.   [*Just Published.* **8/0**
"Represents an enormous quantity of work, and forms a desirable book of reference."—*The Engineer.*
"The volume is distinctly in advance of most similar publications in this country.' — *Engineering.*
"This valuable and well-designed book of reference meets the demands of all descriptions of engineers."—*Saturday Review.*
"Teems with up-to-date information in every branch of engineering and construction. — *Building News.*
"The needs of the engineering profession could hardly be supplied in a more admirable, complete and convenient form. To say that it more than sustains all comparisons is praise of the highest sort, and that may justly be said of it."—*Mining Journal.*
"There is certainly room for the newcomer, which supplies explanations and directions, as well as formulæ and tables. It deserves to become one of the most successful of the technical annuals."—*Architect.*
"Brings together with great skill all the technical information which an engineer has to use day by day. It is in every way admirably equipped, and is sure to prove successful."—*Scotsman.*
"The up-to-dateness of Mr. Kempe's compilation is a quality that will not be lost on the busy people for whom the work is intended."—*Glasgow Herald.*

THE PORTABLE ENGINE.
A Practical Manual on its Construction and Management. For the use of Owners and Users of Steam Engines generally. By WILLIAM DYSON WANSBROUGH. Crown 8vo, cloth **3/6**
"This is a work of value to those who use steam machinery. . . . Should be read by every one who has a steam engine, on a farm or elsewhere."—*Mark Lane Express.*
"We cordially commend this work to buyers and owners of steam-engines, and to those who have to do with their construction or use."—*Timber Trades Journal.*
"Such a general knowledge of the steam-engine as Mr. Wansbrough furnishes to the reader should be acquired by all intelligent owners and others who use the steam-engine."—*Building News.*
"An excellent text-book of this useful form of engine. The 'Hints to Purchasers' contain a good deal of common-sense and practical wisdom."—*English Mechanic.*

IRON AND STEEL.
A Work for the Forge, Foundry, Factory, and Office. Containing ready, useful, and trustworthy Information for Ironmasters and their Stock-takers; Managers of Bar, Rail, Plate, and Sheet Rolling Mills; Iron and Metal Founders; Iron Ship and Bridge Builders; Mechanical, Mining, and Consulting Engineers; Architects, Contractors, Builders, &c. By CHARLES HOARE, Author of "The Slide Rule," &c. Ninth Edition. 32mo, leather . **6/0**
"For comprehensiveness the book has not its equal."—*Iron.*
"One of the best of the pocket books."—*English Mechanic.*

CONDENSED MECHANICS.
A Selection of Formulæ, Rules, Tables, and Data for the Use of Engineering Students, Science Classes, &c. In accordance with the Requirements of the Science and Art Department. By W. G. CRAWFORD HUGHES, A.M.I.C.E. Crown 8vo, cloth **2/6**
"The book is well fitted for those who are either confronted with practical problems in their work, or are preparing for examination and wish to refresh their knowledge by going through their formulæ again."—*Marine Engineer.*
"It is well arranged, and meets the wants of those for whom it is intended."—*Railway News.*

THE SAFE USE OF STEAM.
Containing Rules for Unprofessional Steam Users. By an ENGINEER. Seventh Edition. Sewed. . . . . **6D.**
"If steam-users would but learn this little book by heart, boiler explosions would become sensations by their rarity."—*English Mechanic.*

HEATING BY HOT WATER.
With Information and Suggestions on the best Methods of Heating Public, Private and Horticultural Buildings. By WALTER JONES. Second Edition. With 96 Illustrations, crown 8vo, cloth . . . . . *Net* **2/6**
"We confidently recommend all interested in heating by hot water to secure a copy of this valuable little treatise."—*The Plumber and Decorator.*

CIVIL ENGINEERING, SURVEYING, &c.

LIGHT RAILWAYS FOR THE UNITED KINGDOM, INDIA, AND THE COLONIES.

A Practical Handbook setting forth the Principles on which Light Railways should be Constructed, Worked, and Financed; and detailing the Cost of Construction, Equipment, Revenue and Working Expenses of Local Railways already established in the above-mentioned countries, and in Belgium, France, Switzerland, &c. By J. C. MACKAY, F.G.S., A.M. Inst. C.E. Illustrated with Plates and Diagrams. Medium 8vo, cloth. [*Just Published.* **15/0**

"Mr. Mackay's volume is clearly and concisely written, admirably arranged, and freely illustrated. The book is exactly what has been long wanted. We recommend it to all interested in the subject. It is sure to have a wide sale."—*Railway News.*

"Those who desire to have within reach general information concerning almost all the light railway systems in the world will do well to buy Mr. Mackay's book."—*Engineer.*

"This work appears very opportunely, when the extension of the system on a large scale to England is at last being mooted. In its pages we find all the information that the heart of man can desire on the subject. . . . every detail in its story, founded on the experience of other countries and applied to the possibilities of England, is put before us."—*Spectator.*

PRACTICAL TUNNELLING.

Explaining in detail Setting-out the Works, Shaft-sinking, and Heading-driving, Ranging the Lines and Levelling underground, Sub-Excavating, Timbering and the Construction of the Brickwork of Tunnels, with the amount of Labour required for, and the Cost of, the various portions of the work. By FREDERICK W. SIMMS, M. Inst. C.E. Fourth Edition, Revised and Further Extended, including the most recent (1895) Examples of Sub-aqueous and other Tunnels, by D. KINNEAR CLARK, M. Inst. C.E. Imperial 8vo, with 34 Folding Plates and other Illustrations. Cloth. [*Just Published.* **£2 2s.**

"The present (1896) edition has been brought right up to date, and is thus rendered a work to which civil engineers generally should have ready access, and to which engineers who have construction work can hardly afford to be without, but which to the younger members of the profession is invaluable, as from its pages they can learn the state to which the science of tunnelling has attained."—*Railway News.*

"The estimation in which Mr. Simms's book has been held for many years cannot be more truly expressed than in the words of the late Prof. Rankine: 'The best source of information on the subject of tunnels is Mr. F. W. Simms's work on Practical Tunnelling.'"—*Architect.*

THE WATER SUPPLY OF TOWNS AND THE CONSTRUCTION OF WATER-WORKS.

A Practical Treatise for the Use of Engineers and Students of Engineering. By W. K. BURTON, A.M. Inst. C.E., Professor of Sanitary Engineering in the Imperial University, Tokyo, Japan, and Consulting Engineer to the Tokyo Water-works. Second Edition, Revised and Extended. With numerous Plates and Illustrations. Super-royal 8vo, buckram. [*Just Published.* **25/0**

I. INTRODUCTORY.—II. DIFFERENT QUALITIES OF WATER.—III. QUANTITY OF WATER TO BE PROVIDED.—IV. ON ASCERTAINING WHETHER A PROPOSED SOURCE OF SUPPLY IS SUFFICIENT.—V. ON ESTIMATING THE STORAGE CAPACITY REQUIRED TO BE PROVIDED.—VI. CLASSIFICATION OF WATER-WORKS.—VII. IMPOUNDING RESERVOIRS.—VIII. EARTHWORK DAMS.—IX. MASONRY DAMS.—X. THE PURIFICATION OF WATER.—XI. SETTLING RESERVOIRS.—XII. SAND FILTRATION.—XIII. PURIFICATION OF WATER BY ACTION OF IRON, SOFTENING OF WATER BY ACTION OF LIME, NATURAL FILTRATION.—XIV. SERVICE OR CLEAN WATER RESERVOIRS—WATER TOWERS—STAND PIPES.—XV. THE CONNECTION OF SETTLING RESERVOIRS, FILTER BEDS AND SERVICE RESERVOIRS.—XVI. PUMPING MACHINERY.—XVII. FLOW OF WATER IN CONDUITS—PIPES AND OPEN CHANNELS.—XVIII. DISTRIBUTION SYSTEMS.—XIX. SPECIAL PROVISIONS FOR THE EXTINCTION OF FIRE.—XX. PIPES FOR WATER-WORKS.—XXI. PREVENTION OF WASTE OF WATER.—XXII. VARIOUS APPLICATIONS USED IN CONNECTION WITH WATER-WORKS.

APPENDIX I. By PROF. JOHN MILNE, F.R.S.—CONSIDERATIONS CONCERNING THE PROBABLE EFFECTS OF EARTHQUAKES ON WATER-WORKS, AND THE SPECIAL PRECAUTIONS TO BE TAKEN IN EARTHQUAKE COUNTRIES.

APPENDIX II. By JOHN DE RIJKE, C.E.—ON SAND DUNES AND DUNE SAND AS A SOURCE OF WATER SUPPLY.

"The chapter upon filtration of water is very complete, and the details of construction well illustrated. . . . The work should be specially valuable to civil engineers engaged in work in Japan, but the interest is by no means confined to that locality."—*Engineer.*

"We congratulate the author upon the practical commonsense shown in the preparation of this work. . . . The plates and diagrams have evidently been prepared with great care, and cannot fail to be of great assistance to the student."—*Builder.*

"The whole art of water-works construction is dealt with in a clear and comprehensive fashion in this handsome volume. . . . Mr. Burton's practical treatise shows in all its sections the fruit of independent study and individual experience. It is largely based upon his own practice in the branch of engineering of which it treats."—*Saturday Review.*

THE WATER SUPPLY OF CITIES AND TOWNS.

By WILLIAM HUMBER, A. M. Inst. C.E., and M. Inst. M.E., Author of "Cast and Wrought Iron Bridge Construction," &c., &c. Illustrated with 50 Double Plates, 1 Single Plate, Coloured Frontispiece, and upwards of 250 Woodcuts, and containing 400 pp. of Text. Imp. 4to, elegantly and substantially half-bound in morocco . . . . *Net* **£6 6s.**

LIST OF CONTENTS.

I. HISTORICAL SKETCH OF SOME OF THE MEANS THAT HAVE BEEN ADOPTED FOR THE SUPPLY OF WATER TO CITIES AND TOWNS.—II. WATER AND THE FOREIGN MATTER USUALLY ASSOCIATED WITH IT.—III. RAINFALL AND EVAPORATION.—IV. SPRINGS AND THE WATER-BEARING FORMATIONS OF VARIOUS DISTRICTS.—V. MEASUREMENT AND ESTIMATION OF THE FLOW OF WATER.—VI. ON THE SELECTION OF THE SOURCE OF SUPPLY.—VII. WELLS.—VIII. RESERVOIRS.—IX. THE PURIFICATION OF WATER.—X. PUMPS.—XI. PUMPING MACHINERY.—XII. CONDUITS.—XIII. DISTRIBUTION OF WATER.—XIV. METERS, SERVICE PIPES, AND HOUSE FITTINGS.—XV. THE LAW OF ECONOMY OF WATER-WORKS.—XVI. CONSTANT AND INTERMITTENT SUPPLY.—XVII. DESCRIPTION OF PLATES.—APPENDICES, GIVING TABLES OF RATES OF ¦SUPPLY, VELOCITIES, &c., &c., TOGETHER WITH SPECIFICATIONS OF SEVERAL WORKS ILLUSTRATED, AMONG WHICH WILL BE FOUND: ABERDEEN, BIDEFORD, CANTERBURY, DUNDEE, HALIFAX, LAMBETH, ROTHERHAM, DUBLIN, AND OTHERS.

"The most systematic and valuable work upon water supply hitherto produced in English, or in any other language. . . . Mr. Humber's work is characterised almost throughout by an exhaustiveness much more distinctive of French and German than of English technical treatises."—*Engineer.*

RURAL WATER SUPPLY.

A Practical Handbook on the Supply of Water and Construction of Waterworks for small Country Districts. By ALLAN GREENWELL, A.M.I.C.E., and W. T. CURRY, A.M.I.C.E., F.G.S. With Illustrations. Second Edition, Revised. Crown 8vo, cloth. [*Just Published.* **5/0**

"We conscientiously recommend it as a very useful book for those concerned in obtaining water for small districts, giving a great deal of practical information in a small compass."—*Builder.*
"The volume contains valuable information upon all matters connected with water supply . . . It is full of details on points which are continually before water-works engineers."—*Nature.*

HYDRAULIC TABLES, CO-EFFICIENTS, & FORMULÆ.

For Finding the Discharge of Water from Orifices, Notches, Weirs, Pipes, and Rivers. With New Formulæ, Tables, and General Information on Rain-fall, Catchment-Basins, Drainage, Sewerage, Water Supply for Towns and Mill Power. By JOHN NEVILLE, Civil Engineer, M.R.I.A. Third Edition, revised, with additions. Numerous Illustrations. Crown 8vo, cloth . **14/0**

"It is, of all English books on the subject, the one nearest to completeness. . . . From the good arrangement of the matter, the clear explanations and abundance of formulæ, the carefully calculated tables, and, above all, the thorough acquaintance with both theory and construction, which is displayed from first to last, the book will be found to be an acquisition."—*Architect.*

HYDRAULIC MANUAL.

Consisting of Working Tables and Explanatory Text. Intended as a Guide in Hydraulic Calculations and Field Operations. By LOWIS D'A. JACKSON, Author of "Aid to Survey Practice," "Modern Metrology," &c. Fourth Edition, Enlarged. Large crown 8vo, cloth **16/0**

"The author has had a wide experience in hydraulic engineering and has been a careful observer of the facts which have come under his notice, and from the great mass of material at his command he has constructed a manual which may be accepted as a trustworthy guide to this branch of the engineer's profession."—*Engineering.*
"The most useful feature of this work is its freedom from what is superannuated, and its thorough adoption of recent experiments; the text is in fact in great part a short account of the great modern experiments."—*Nature.*

WATER ENGINEERING.

A Practical Treatise on the Measurement, Storage, Conveyance, and Utilisation of Water for the Supply of Towns, for Mill Power, and for other Purposes. By C. SLAGG, A. M. Inst. C.E. Second Edition. Crown 8vo, cloth . **7/6**

"As a small practical treatise on the water supply of towns, and on some applications of waterpower, the work is in many respects excellent."—*Engineering.*
"The author has collated the results deduced from the experiments of the most eminent authorities, and has presented them in a compact and practical form, accompanied by very clear and detailed explanations. . . . The application of water as a motive power is treated very carefully and exhaustively."—*Builder.*

MASONRY DAMS FROM INCEPTION TO COMPLETION.

Including numerous Formulæ, Forms of Specification and Tender, Pocket Diagram of Forces, &c. For the use of Civil and Mining Engineers. By C. F. COURTNEY, M. Inst. C.E. 8vo, cloth. [*Just Published.* **9/0**

RIVER BARS.

The Causes of their Formation, and their Treatment by "Induced Tidal Scour;" with a Description of the Successful Reduction by this Method of the Bar at Dublin. By I. J. MANN, Assist. Eng. to the Dublin Port and Docks Board. Royal 8vo, cloth **7/6**

"We recommend all interested in harbour works—and, indeed, those concerned in the improvements of rivers generally—to read Mr. Mann's interesting work on the treatment of river bars."—*Engineer.*

DRAINAGE OF LANDS, TOWNS AND BUILDINGS.

By G. D. DEMPSEY, C.E. Revised, with large Additions on RECENT PRACTICE IN DRAINAGE ENGINEERING, by D. KINNEAR CLARK, M. Inst. C.E., Author of "Tramways: their Construction and Working." Cr. 8vo, cloth . **4/6**

"The new matter added to Mr. Dempsey's excellent work is characterised by the comprehensive grasp and accuracy of detail for which the name of Mr. D. K. Clark is a sufficient voucher."—*Athenæum.*

TRAMWAYS: THEIR CONSTRUCTION AND WORKING.

Embracing a Comprehensive History of the System; with an exhaustive Analysis of the Various Modes of Traction, including Horse Power, Steam, Cable Traction, Electric Traction, &c.; a Description of the Varieties of Rolling Stock; and ample Details of Cost and Working Expenses. New Edition, Thoroughly Revised, and Including the Progress recently made in Tramway Construction, &c., &c. By D. KINNEAR CLARK, M. Inst. C.E. With 400 Illustrations. 8vo, 780 pp., buckram. [*Just Published.* **28/0**

"Although described as a new edition, this book is really a new one, a large part of it, which covers historical ground, having been re-written and amplified; while the parts which relate to all that has been done since 1882 appear in this edition only. It is sixteen years since the first edition appeared, and twelve years since the supplementary volume to the first book was published. After a lapse, then, of twelve years, it is obvious that the author has at his disposal a vast quantity of descriptive and statistical information, with which he may, and has, produced a volume of great value to all interested in tramway construction and working. The new volume is one which will rank, among tramway engineers and those interested in tramway working, with his world-famed book on railway machinery."—*The Engineer*, March 8, 1895.

PRACTICAL SURVEYING.

A Text-Book for Students preparing for Examinations or for Survey-work in the Colonies. By GEORGE W. USILL, A.M.I.C.E. With 4 Plates and upwards of 330 Illustrations. Fifth Edition, Revised and Enlarged. Including Tables of Natural Sines, Tangents, Secants, &c. Crown 8vo, cloth **7/6**; or, on THIN PAPER, bound in limp leather, gilt edges, rounded corners, for pocket use **12/6**

"The best forms of instruments are described as to their construction, uses and modes of employment, and there are innumerable hints on work and equipment such as the author, in his experience as surveyor, draughtsman and teacher, has found necessary, and which the student in his inexperience will find most serviceable."—*Engineer.*

"The latest treatise in the English language on surveying, and we have no hesitation in saying that the student will find it a better guide than any of its predecessors. Deserves to be recognised as the first book which should be put in the hands of a pupil of Civil Engineering."—*Architect.*

AID TO SURVEY PRACTICE.

For Reference in Surveying, Levelling, and Setting-out; and in Route Surveys of Travellers by Land and Sea. With Tables, Illustrations, and Records. By LOWIS D'A. JACKSON, A.M.I.C.E. Second Edition, Enlarged. Large crown 8vo, cloth **12/6**

"Mr. Jackson has produced a valuable *vade-mecum* for the surveyor. We can recommend this book as containing an admirable supplement to the teaching of the accomplished surveyor."—*Athenæum.*

"As a text-book we should advise all surveyors to place it in their libraries, and study well the matured instructions afforded in its pages."—*Colliery Guardian.*

"The author brings to his work a fortunate union of theory and practical experience which, aided by a clear and lucid style of writing, renders the book a very useful one."—*Builder.*

12   CROSBY LOCKWOOD & SON'S CATALOGUE.

ENGINEER'S & MINING SURVEYOR'S FIELD BOOK.

Consisting of a Series of Tables, with Rules, Explanations of Systems, and use of Theodolite for Traverse Surveying and plotting the work with minute accuracy by means of Straight Edge and Set Square only; Levelling with the Theodolite, Casting-out and Reducing Levels to Datum, and Plotting Sections in the ordinary manner; Setting-out Curves with the Theodolite by Tangential Angles and Multiples with Right and Left-hand Readings of the Instrument; Setting-out Curves without Theodolite on the System of Tangential Angles by Sets of Tangents and Offsets; and Earthwork Tables to 80 feet deep, calculated for every 6 inches in depth. By W. DAVIS HASKOLL, C.E. With numerous Woodcuts. Fourth Edition, Enlarged. Crown 8vo, cloth . **12/0**

"The book is very handy; the separate tables of sines and tangents to every minute will make it useful for many other purposes, the genuine traverse tables existing all the same."—*Athenæum.*
"Every person engaged in engineering field operations will estimate the importance of such a work and the amount of valuable time which will be saved by reference to a set of reliable tables prepared with the accuracy and fulness of those given in this volume."—*Railway News.*

LAND AND MARINE SURVEYING.

In Reference to the Preparation of Plans for Roads and Railways; Canals, Rivers, Towns' Water Supplies; Docks and Harbours. With Description and Use of Surveying Instruments. By W. DAVIS HASKOLL, C.E. Second Edition, Revised, with Additions. Large crown 8vo, cloth . . . **9/0**

"This book must prove of great value to the student. We have no hesitation in recommending it, feeling assured that it will more than repay a careful study."—*Mechanical World.*
"A most useful book for the student. We strongly recommend it as a carefully-written and valuable text-book. It enjoys a well-deserved repute among surveyors."—*Builder.*
"This volume cannot fail to prove of the utmost practical utility. It may be safely recommended to all students who aspire to become clean and expert surveyors."—*Mining Journal.*

PRINCIPLES AND PRACTICE OF LEVELLING.

Showing its Application to Purposes of Railway and Civil Engineering in the Construction of Roads; with Mr. TELFORD'S Rules for the same. By FREDERICK W. SIMMS, F.G.S., M. Inst. C.E. Eighth Edition, with the addition of LAW'S Practical Examples for Setting-out Railway Curves, and TRAUTWINE'S Field Practice of Laying-out Circular Curves. With 7 Plates and numerous Woodcuts, 8vo, cloth **8/6**
*** TRAUTWINE on CURVES may be had separate . . . **5/0**

"The text-book on levelling in most of our engineering schools and colleges."—*Engineer.*
"The publishers have rendered a substantial service to the profession, especially to the younger members, by bringing out the present edition of Mr. Simms's useful work."—*Engineering.*

AN OUTLINE OF THE METHOD OF CONDUCTING A TRIGONOMETRICAL SURVEY.

For the Formation of Geographical and Topographical Maps and Plans, Military Reconnaissance, LEVELLING, &c., with Useful Problems, Formulæ, and Tables. By Lieut.-General FROME, R.E. Fourth Edition, Revised and partly Re-written by Major-General Sir CHARLES WARREN, G.C.M.G., R.E. With 19 Plates and 115 Woodcuts, royal 8vo, cloth . . . **16/0**

"No words of praise from us can strengthen the position so well and so steadily maintained by this work. Sir Charles Warren has revised the entire work, and made such additions as were necessary to bring every portion of the contents up to the present date."—*Broad Arrow.*

TABLES OF TANGENTIAL ANGLES AND MULTIPLES FOR SETTING-OUT CURVES.

From 5 to 200 Radius. By A. BEAZELEY, M. Inst. C.E. 6th Edition, Revised. With an Appendix on the use of the Tables for Measuring up Curves. Printed on 50 Cards, and sold in a cloth box, waistcoat-pocket size.
[*Just Published.* **3/6**
"Each table is printed on a card, which, placed on the theodolite, leaves the hands free to manipulate the instrument—no small advantage as regards the rapidity of work."—*Engineer.*
"Very handy: a man may know that all his day's work must fall on two of these cards, which he puts into his own card-case, and leaves the rest behind."—*Athenæum.*

HANDY GENERAL EARTH-WORK TABLES.

Giving the Contents in Cubic Yards of Centre and Slopes of Cuttings and Embankments from 3 inches to 80 feet in Depth or Height, for use with either 66 feet Chain or 100 feet Chain. By J. H. WATSON BUCK, M. Inst. C.E. On a Sheet mounted in cloth case. [*Just Published.* **3/6**

CIVIL ENGINEERING, SURVEYING, &c.   13

EARTHWORK TABLES.
Showing the Contents in Cubic Yards of Embankments, Cuttings, &c., of Heights or Depths up to an average of 80 feet. By JOSEPH BROADBENT, C.E., and FRANCIS CAMPIN, C.E. Crown 8vo, cloth **5/0**
"The way in which accuracy is attained, by a simple division of each cross section into three elements, two in which are constant and one variable, is ingenious."—*Athenæum*.

A MANUAL ON EARTHWORK.
By ALEX. J. S. GRAHAM, C.E. With numerous Diagrams. Second Edition. 18mo, cloth **2/6**

THE CONSTRUCTION OF LARGE TUNNEL SHAFTS.
A Practical and Theoretical Essay. By J. H. WATSON BUCK, M. Inst. C.E., Resident Engineer, L. and N. W. R. With Folding Plates, 8vo, cloth **12/0**
"Many of the methods given are of extreme practical value to the mason, and the observations on the form of arch, the rules for ordering the stone, and the construction of the templates, will be found of considerable use. We commend the book to the engineering profession."—*Building News*.
"Will be regarded by civil engineers as of the utmost value, and calculated to save much time and obviate many mistakes."—*Colliery Guardian*.

CAST & WROUGHT IRON BRIDGE CONSTRUCTION.
(A Complete and Practical Treatise on), including Iron Foundations. In Three Parts.—Theoretical, Practical, and Descriptive. By WILLIAM HUMBER, A. M. Inst. C.E., and M. Inst. M.E. Third Edition, revised and much improved, with 115 Double Plates (20 of which now first appear in this edition), and numerous Additions to the Text. In 2 vols., imp. 4to, half-bound in morocco . . . . **£6 16s. 6d.**
"A very valuable contribution to the standard literature of civil engineering. In addition to elevations, plans, and sections, large scale details are given, which very much enhance the instructive worth of those illustrations."—*Civil Engineer and Architect's Journal*.
"Mr. Humber's stately volumes, lately issued—in which the most important bridges erected during the last five years, under the direction of the late Mr. Brunel, Sir W. Cubitt, Mr. Hawkshaw, Mr. Page, Mr. Fowler, Mr. Hemans, and others among our most eminent engineers, are drawn and specified in great detail."—*Engineer*.

ESSAY ON OBLIQUE BRIDGES
(Practical and Theoretical). With 13 large Plates. By the late GEORGE WATSON BUCK, M.I.C.E. Fourth Edition, revised by his Son, J. H. WATSON BUCK, M.I.C.E.; and with the addition of Description to Diagrams for Facilitating the Construction of Oblique Bridges, by W. H. BARLOW, M.I.C.E. Royal 8vo, cloth **12/0**
"The standard text-book for all engineers regarding skew arches is Mr. Buck's treatise, and it would be impossible to consult a better."—*Engineer*.
"Mr. Buck's treatise is recognised as a standard text-book, and his treatment has divested the subject of many of the intricacies supposed to belong to it. As a guide to the engineer and architect, on a confessedly difficult subject, Mr. Buck's work is unsurpassed."—*Building News*.

THE CONSTRUCTION OF OBLIQUE ARCHES
(A Practical Treatise on). By JOHN HART. Third Edition, with Plates. Imperial 8vo, cloth **8/0**

GRAPHIC AND ANALYTIC STATICS.
In their Practical Application to the Treatment of Stresses in Roofs, Solid Girders, Lattice, Bowstring, and Suspension Bridges, Braced Iron Arches and Piers, and other Frameworks. By R. HUDSON GRAHAM, C.E. Containing Diagrams and Plates to Scale. With numerous Examples, many taken from existing Structures. Specially arranged for Class-work in Colleges and Universities. Second Edition, Revised and Enlarged. 8vo, cloth . **16/0**
"Mr. Graham's book will find a place wherever graphic and analytic statics are used or studied."—*Engineer*.
"The work is excellent from a practical point of view, and has evidently been prepared with much care. The directions for working are simple, and are illustrated by an abundance of well-selected examples. It is an excellent text-book for the practical draughtsman."—*Athenæum*.

PRACTICAL GEOMETRY.
For the Architect, Engineer, and Mechanic. Giving Rules for the Delineation and Application of various Geometrical Lines, Figures, and Curves. By E. W. TARN, M.A., Architect. 8vo, cloth **9/0**
"No book with the same objects in view has ever been published in which the clearness of the rules laid down and the illustrative diagrams have been so satisfactory."—*Scotsman*.

THE GEOMETRY OF COMPASSES.
Or, Problems Resolved by the mere Description of Circles and the Use of Coloured Diagrams and Symbols. By OLIVER BYRNE. Coloured Plates. Crown 8vo, cloth **3/6**

WEIGHTS OF WROUGHT IRON & STEEL GIRDERS.
A Graphic Table for Facilitating the Computation of the Weights of Wrought Iron and Steel Girders, &c., for Parliamentary and other Estimates. By J. H. WATSON BUCK, M. Inst. C.E. On a Sheet **2/6**

HANDY BOOK FOR THE CALCULATION OF STRAINS
In Girders and Similar Structures and their Strength. Consisting of Formulæ and Corresponding Diagrams, with numerous details for Practical Application, &c. By WILLIAM HUMBER, A. M. Inst. C.E., &c. Fifth Edition. Crown 8vo, with nearly 100 Woodcuts and 3 Plates, cloth . . . **7/6**
"The formulæ are neatly expressed, and the diagrams good."—*Athenæum.*
"We heartily commend this really *handy* book to our engineer and architect readers."—*English Mechanic.*

TRUSSES OF WOOD AND IRON.
Practical Applications of Science in Determining the Stresses, Breaking Weights, Safe Loads, Scantlings, and Details of Construction. With Complete Working Drawings. By WILLIAM GRIFFITHS, Surveyor, Assistant Master, Tranmere School of Science and Art. Oblong 8vo, cloth . . . **4/6**
"This handy little book enters so minutely into every detail connected with the construction of roof trusses that no student need be ignorant of these matters."—*Practical Engineer.*

THE STRAINS ON STRUCTURES OF IRONWORK.
With Practical Remarks on Iron Construction. By F. W. SHEILDS, M.I.C.E. 8vo, cloth **5/0**

A TREATISE ON THE STRENGTH OF MATERIALS.
With Rules for Application in Architecture, the Construction of Suspension Bridges, Railways, &c. By PETER BARLOW, F.R.S. A new Edition, revised by his Sons, P. W. BARLOW, F.R.S., and W. H. BARLOW, F.R.S.; to which are added, Experiments by HODGKINSON, FAIRBAIRN, and KIRKALDY; and Formulæ for calculating Girders, &c. Arranged and Edited by WM. HUMBER, A. M. Inst. C.E. Demy 8vo, 400 pp., with 19 large Plates and numerous Woodcuts, cloth **18/0**
"Valuable alike to the student, tyro, and the experienced practitioner, it will always rank in future as it has hitherto done, as the standard treatise on that particular subject."—*Engineer.*
"As a scientific work of the first class, it deserves a foremost place on the bookshelves of every civil engineer and practical mechanic."—*English Mechanic.*

STRENGTH OF CAST IRON AND OTHER METALS.
By THOMAS TREDGOLD, C.E. Fifth Edition, including HODGKINSON'S Experimental Researches. 8vo, cloth **12/0**

SAFE RAILWAY WORKING.
A Treatise on Railway Accidents, their Cause and Prevention; with a Description of Modern Appliances and Systems. By CLEMENT E. STRETTON, C.E., Vice-President and Consulting Engineer, Amalgamated Society of Railway Servants. With Illustrations and Coloured Plates. Third Edition, Enlarged. Crown 8vo, cloth **3/6**
"A book for the engineer, the directors, the managers; and, in short, all who wish for information on railway matters will find a perfect encyclopædia in 'Safe Railway Working.'"—*Railway Review.*
"We commend the remarks on railway signalling to all railway managers, especially where a uniform code and practice is advocated."—*Herepath's Railway Journal.*

EXPANSION OF STRUCTURES BY HEAT.
By JOHN KEILY, C.E., late of the Indian Public Works Department. Crown 8vo, cloth **3/6**
"The aim the author has set before him, viz., to show the effects of heat upon metallic and other structures, is a laudable one, for this is a branch of physics upon which the engineer or architect can find but little reliable and comprehensive data in books."—*Builder.*

CIVIL ENGINEERING, SURVEYING, &c. 15

RECORD OF THE PROGRESS OF MODERN ENGINEERING.

Complete in Four Volumes, imperial 4to, half-morocco, price **£12 12s.** Each volume sold separately, as follows:—

FIRST SERIES, Comprising Civil, Mechanical, Marine, Hydraulic, Railway, Bridge, and other Engineering Works, &c. By WILLIAM HUMBER, A. M. Inst. C.E., &c. Imp. 4to, with 36 Double Plates, drawn to a large scale, Photographic Portrait of John Hawkshaw, C.E., F.R.S., &c., and copious descriptive Letterpress, Specifications, &c. Half-morocco . . **£3 3s.**

LIST OF THE PLATES AND DIAGRAMS.

VICTORIA STATION AND ROOF, L. B. & S. C. R. (8 PLATES); SOUTHPORT PIER (2 PLATES); VICTORIA STATION AND ROOF, L. C. & D. AND G. W. R. (6 PLATES); ROOF OF CREMORNE MUSIC HALL; BRIDGE OVER G. N. RAILWAY; ROOF OF STATION, DUTCH RHENISH RAIL. (2 PLATES); BRIDGE OVER THE THAMES, WEST LONDON EXTENSION RAILWAY (5 PLATES); ARMOUR PLATES; SUSPENSION BRIDGE, THAMES (4 PLATES); THE ALLEN ENGINE; SUSPENSION BRIDGE, AVON (3 PLATES); UNDERGROUND RAILWAY (3 PLATES).

HUMBER'S PROGRESS OF MODERN ENGINEERING.

SECOND SERIES. Imp. 4to, with 3 Double Plates, Photographic Portrait of Robert Stephenson, C.E., M.P., F.R.S., &c., and copious descriptive Letterpress, Specifications, &c. Half-morocco . . . . . . **£3 3s.**

LIST OF THE PLATES AND DIAGRAMS.

BIRKENHEAD DOCKS, LOW WATER BASIN (15 PLATES); CHARING CROSS STATION ROOF, C. C. RAILWAY (3 PLATES); DIGSWELL VIADUCT, GREAT NORTHERN RAILWAY; ROBBERY WOOD VIADUCT, GREAT NORTHERN RAILWAY; IRON PERMANENT WAY; CLYDACH VIADUCT, MERTHYR, TREDEGAR, AND ABERGAVENNY RAILWAY; EBBW VIADUCT, MERTHYR, TREDEGAR, AND ABERGAVENNY RAILWAY; COLLEGE WOOD VIADUCT, CORNWALL RAILWAY; DUBLIN WINTER PALACE ROOF (3 PLATES); BRIDGE OVER THE THAMES, L. C. & D. RAILWAY (6 PLATES); ALBERT HARBOUR, GREENOCK (4 PLATES).

HUMBER'S PROGRESS OF MODERN ENGINEERING.

THIRD SERIES. Imp. 4to, with 40 Double Plates, Photographic Portrait of J. R. M'Clean, late Pres. Inst. C.E., and copious descriptive Letterpress, Specifications, &c. Half-morocco . . . . **£3 3s.**

LIST OF THE PLATES AND DIAGRAMS.

MAIN DRAINAGE, METROPOLIS.—*North Side.*—MAP SHOWING INTERCEPTION OF SEWERS; MIDDLE LEVEL SEWER (2 PLATES); OUTFALL SEWER, BRIDGE OVER RIVER LEA (3 PLATES); OUTFALL SEWER, BRIDGE OVER MARSH LANE, NORTH WOOLWICH RAILWAY, AND BOW AND BARKING RAILWAY JUNCTION; OUTFALL SEWER, BRIDGE OVER BOW AND BARKING RAILWAY (3 PLATES); OUTFALL SEWER, BRIDGE OVER EAST LONDON WATER-WORKS' FEEDER (2 PLATES); OUTFALL SEWER RESERVOIR (2 PLATES); OUTFALL SEWER, TUMBLING BAY AND OUTLET; OUTFALL SEWER, PENSTOCKS. *South Side.*— OUTFALL SEWER, BERMONDSEY BRANCH (2 PLATES); OUTFALL SEWER, RESERVOIR AND OUTLET (4 PLATES); OUTFALL SEWER, FILTH HOIST; SECTIONS OF SEWERS (NORTH AND SOUTH SIDES). THAMES EMBANKMENT.—SECTION OF RIVER WALL; STEAMBOAT PIER, WESTMINSTER (2 PLATES); LANDING STAIRS BETWEEN CHARING CROSS AND WATERLOO BRIDGES; YORK GATE (2 PLATES); OVERFLOW AND OUTLET AT SAVOY STREET SEWER (3 PLATES); STEAMBOAT PIER, WATERLOO BRIDGE (3 PLATES); JUNCTION OF SEWERS, PLANS AND SECTIONS; GULLIES, PLANS AND SECTIONS; ROLLING STOCK; GRANITE AND IRON FORTS.

HUMBER'S PROGRESS OF MODERN ENGINEERING.

FOURTH SERIES. Imp. 4to, with 36 Double Plates, Photographic Portrait of John Fowler, late Pres. Inst. C.E., and copious descriptive Letterpress, Specifications, &c. Half-morocco . . . . **£3 3s.**

LIST OF THE PLATES AND DIAGRAMS.

ABBEY MILLS PUMPING STATION, MAIN DRAINAGE, METROPOLIS (4 PLATES); BARROW DOCKS (5 PLATES); MANQUIS VIADUCT, SANTIAGO AND VALPARAISO RAILWAY, (2 PLATES); ADAM'S LOCOMOTIVE, ST. HELEN'S CANAL RAILWAY (2 PLATES); CANNON STREET STATION ROOF, CHARING CROSS RAILWAY (3 PLATES); ROAD BRIDGE OVER THE RIVER MOKA (2 PLATES); TELEGRAPHIC APPARATUS FOR MESOPOTAMIA; VIADUCT OVER THE RIVER WYE, MIDLAND RAILWAY (3 PLATES); ST. GERMANS VIADUCT, CORNWALL RAILWAY (2 PLATES); WROUGHT-IRON CYLINDER FOR DIVING BELL; MILLWALL DOCKS (6 PLATES); MILROY'S PATENT EXCAVATOR; METROPOLITAN DISTRICT RAILWAY (6 PLATES); HARBOURS, PORTS, AND BREAKWATERS (3 PLATES).

16   CROSBY LOCKWOOD & SON'S CATALOGUE.

THE POPULAR WORKS OF MICHAEL REYNOLDS.

LOCOMOTIVE ENGINE DRIVING.

A Practical Manual for Engineers in Charge of Locomotive Engines. By MICHAEL REYNOLDS, Member of the Society of Engineers, formerly Locomotive Inspector, L. B. & S. C. R. Ninth Edition. Including a KEY TO THE LOCOMOTIVE ENGINE. With Illustrations and Portrait of Author. Crown 8vo, cloth **4/6**

"Mr. Reynolds has supplied a want, and has supplied it well. We can confidently recommend the book not only to the practical driver, but to everyone who takes an interest in the performance of locomotive engines."—*The Engineer.*
"Mr. Reynolds has opened a new chapter in the literature of the day. This admirable practical treatise, of the practical utility of which we have to speak in terms of warm commendation.' —*Athenæum.*
"Evidently the work of one who knows his subject thoroughly."—*Railway Service Gazette.*
"Were the cautions and rules given in the book to become part of the every-day working of our engine-drivers, we might have fewer distressing accidents to deplore."—*Scotsman.*

STATIONARY ENGINE DRIVING.

A Practical Manual for Engineers in Charge of Stationary Engines. By MICHAEL REYNOLDS. Fifth Edition, Enlarged. With Plates and Woodcuts. Crown 8vo, cloth **4/6**

"The author is thoroughly acquainted with his subjects, and his advice on the various points treated is clear and practical. . . . He has produced a manual which is an exceedingly useful one for the class for whom it is specially intended."—*Engineering.*
"Our author leaves no stone unturned. He is determined that his readers shall not only know something about the stationary engine, but all about it."—*Engineer.*
"An engineman who has mastered the contents of Mr. Reynolds's book will require but little actual experience with boilers and engines before he can be trusted to look after them."— *English Mechanic.*

THE MODEL LOCOMOTIVE ENGINEER,

Fireman, and Engine-Boy. Comprising a Historical Notice of the Pioneer Locomotive Engines and their Inventors. By MICHAEL REYNOLDS. Second Edition, with Revised Appendix. With numerous Illustrations, and Portrait of George Stephenson. Crown 8vo, cloth. [*Just Published.* **4/6**

"From the technical knowledge of the author, it will appeal to the railway man of to-day more forcibly than anything written by Dr. Smiles. . . . The volume contains information of a technical kind, and facts that every driver should be familiar with."—*English Mechanic.*
"We should be glad to see this book in the possession of everyone in the kingdom who has ever laid, or is to lay, hands on a locomotive engine."—*Iron.*

CONTINUOUS RAILWAY BRAKES.

A Practical Treatise on the several Systems in Use in the United Kingdom : their Construction and Performance. With copious Illustrations and numerous Tables. By MICHAEL REYNOLDS. Large crown 8vo, cloth . . . **9/0**

"A popular explanation of the different brakes. It will be of great assistance in forming public opinion, and will be studied with benefit by those who take an interest in the brake."—*English Mechanic.*
"Written with sufficient technical detail to enable the principal and relative connection of the various parts of each particular brake to be readily grasped."—*Mechanical World.*

ENGINE-DRIVING LIFE.

Stirring Adventures and Incidents in the Lives of Locomotive Engine-Drivers. By MICHAEL REYNOLDS. Third Edition. Crown 8vo, cloth . **1/6**

"From first to last perfectly fascinating. Wilkie Collins's most thrilling conceptions are thrown into the shade by true incidents, endless in their variety, related in every page."—*North British Mail.*
"Anyone who wishes to get a real insight into railway fe cannot do better than read 'Engine-Driving Life' for himself, and if he once takes it up he will find that the author's enthusiasm and real love of the engine-driving profession will carry him on until he has read every page." —*Saturday Review.*

THE ENGINEMAN'S POCKET COMPANION,

And Practical Educator for Enginemen, Boiler Attendants, and Mechanics By MICHAEL REYNOLDS. With 45 Illustrations and numerous Diagrams. Third Edition, Revised. Royal 18mo, strongly bound for pocket wear . **3/6**

"This admirable work is well suited to accomplish its object, being the honest workmanship of a competent engineer."—*Glasgow Herald.*
"A most meritorious work, giving in a succinct and practical form all the information an engine-minder desirous of mastering the scientific principles of his daily calling would require.' — *The Miller.*
"A boon to those who are striving to become efficient mechanics."—*Daily Chronicle.*

MARINE ENGINEERING, SHIPBUILDING, NAVIGATION, &c.

THE NAVAL ARCHITECT'S AND SHIPBUILDER'S POCKET-BOOK of Formulæ, Rules, and Tables, and Marine Engineer's and Surveyor's Handy Book of Reference. By CLEMENT MACKROW, M.I.N.A. Sixth Edition, Revised, 700 pp., with 300 Illustrations. Fcap., leather. **12/6**

SUMMARY OF CONTENTS:—SIGNS AND SYMBOLS, DECIMAL FRACTIONS.—TRIGONOMETRY.—PRACTICAL GEOMETRY.—MENSURATION.—CENTRES AND MOMENTS OF FIGURES.—MOMENTS OF INERTIA AND RADII OF GYRATION.—ALGEBRAICAL EXPRESSIONS FOR SIMPSON'S RULES.—MECHANICAL PRINCIPLES.—CENTRE OF GRAVITY.—LAWS OF MOTION.—DISPLACEMENT, CENTRE OF BUOYANCY.—CENTRE OF GRAVITY OF SHIP'S HULL.—STABILITY CURVES AND METACENTRES.—SEA AND SHALLOW-WATER WAVES.—ROLLING OF SHIPS.—PROPULSION AND RESISTANCE OF VESSELS.—SPEED TRIALS.—SAILING, CENTRE OF EFFORT.—DISTANCES DOWN RIVERS, COAST LINES.—STEERING AND RUDDERS OF VESSELS.—LAUNCHING CALCULATIONS AND VELOCITIES.—WEIGHT OF MATERIAL AND GEAR.—GUN PARTICULARS AND WEIGHT.—STANDARD GAUGES.—RIVETED JOINTS AND RIVETING.—STRENGTH AND TESTS OF MATERIALS.—BINDING AND SHEARING STRESSES, &c.—STRENGTH OF SHAFTING, PILLARS, WHEELS, &c.—HYDRAULIC DATA, &c.—CONIC SECTIONS, CATENARIAN CURVES.—MECHANICAL POWERS, WORK.—BOARD OF TRADE REGULATIONS FOR BOILERS AND ENGINES.—BOARD OF TRADE REGULATIONS FOR SHIPS.—LLOYD'S RULES FOR BOILERS.—LLOYD'S WEIGHT OF CHAINS.—LLOYD'S SCANTLINGS FOR SHIPS.—DATA OF ENGINES AND VESSELS.—SHIPS' FITTINGS AND TESTS.—SEASONING PRESERVING TIMBER.—MEASUREMENT OF TIMBER.—ALLOYS, PAINTS, VARNISHES.—DATA FOR STOWAGE.—ADMIRALTY TRANSPORT REGULATIONS.—RULES FOR HORSE-POWER, SCREW PROPELLERS, &c.—PERCENTAGES FOR BUTT STRAPS, &c.—PARTICULARS OF YACHTS.—MASTING AND RIGGING VESSELS.—DISTANCES OF FOREIGN PORTS.—TONNAGE TABLES.—VOCABULARY OF FRENCH AND ENGLISH TERMS.—ENGLISH WEIGHTS AND MEASURES.—FOREIGN WEIGHTS AND MEASURES.—DECIMAL EQUIVALENTS.—FOREIGN MONEY.—DISCOUNT AND WAGES TABLES.—USEFUL NUMBERS AND READY RECKONERS.—TABLES OF CIRCULAR MEASURES.—TABLES OF AREAS OF AND CIRCUMFERENCES OF CIRCLES.—TABLES OF AREAS OF SEGMENTS OF CIRCLES.—TABLES OF SQUARES AND CUBES AND ROOTS OF NUMBERS.—TABLES OF LOGARITHMS OF NUMBERS.—TABLES OF HYPERBOLIC LOGARITHMS.—TABLES OF NATURAL SINES, TANGENTS, &c.—TABLES OF LOGARITHMIC SINES, TANGENTS, &c.

"In these days of advanced knowledge a work like this is of the greatest value. It contains a vast amount of information. We unhesitatingly say that it is the most valuable compilation for its specific purpose that has ever been printed. No naval architect, engineer, surveyor, or seaman, wood or iron shipbuilder, can afford to be without this work."—*Nautical Magazine.*

"Should be used by all who are engaged in the construction or design of vessels. . . . Will be found to contain the most useful tables and formulæ required by shipbuilders, carefully collected from the best authorities, and put together in a popular and simple form. The book is one of exceptional merit."—*Engineer.*

"The professional shipbuilder has now, in a convenient and accessible form, reliable data for solving many of the numerous problems that present themselves in the course of his work."—*Iron.*

"There is no doubt that a pocket-book of this description must be a necessity in the shipbuilding trade. . . . The volume contains a mass of useful information clearly expressed and presented in a handy form."—*Marine Engineer.*

WANNAN'S MARINE ENGINEER'S GUIDE
To Board of Trade Examinations for Certificates of Competency. Containing all Latest Questions to Date, with Simple, Clear, and Correct Solutions; Elementary and Verbal Questions and Answers; complete Set of Drawings with Statements completed. By A. C. WANNAN, C.E., and E. W. I. WANNAN, M.I.M.E. Illustrated with numerous Engravings. Crown 8vo, 370 pages, cloth. [*Just Published.* **8/6**

WANNAN'S MARINE ENGINEER'S POCKET-BOOK.
Containing the Latest Board of Trade Rules and Data for Marine Engineers. By A. C. WANNAN. Second Edition, carefully Revised. Square 18mo, with thumb Index, leather. [*Just Published.* **5/0**

MARINE ENGINES AND STEAM VESSELS.
A Treatise on. By ROBERT MURRAY, C.E. Eighth Edition, thoroughly Revised, with considerable Additions by the Author and by GEORGE CARLISLE, C.E., Senior Surveyor to the Board of Trade. 12mo, cloth. **4/6**

"Well adapted to give the young steamship engineer or marine engine and boiler maker a general introduction into his practical work."—*Mechanical World.*

"We feel sure that this thoroughly revised edition will continue to be as popular in the future as it has been in the past, as, for its size, it contains more useful information than any similar treatise."—*Industries.*

SEA TERMS, PHRASES, AND WORDS

(Technical Dictionary of) used in the English and French Languages (English-French, French-English). For the Use of Seamen, Engineers, Pilots, Shipbuilders, Shipowners, and Ship-brokers. Compiled by W. PIRRIE, late of the African Steamship Company. Fcap. 8vo, cloth limp. . . . **5/0**

"This volume will be highly appreciated by seamen, engineers, pilots, shipbuilders and ship-owners. It will be found wonderfully accurate and complete."—*Scotsman.*
"A very useful dictionary, which has long been wanted by French and English engineers, masters, officers and others."—*Shipping World.*

ELECTRIC SHIP LIGHTING.

A Handbook on the Practical Fitting and Running of Ships' Electrical Plant, for the Use of Shipowners and Builders, Marine Electricians and Sea-going Engineers in Charge. By J. W. URQUHART, Author of "Electric Light," "Dynamo Construction," &c. Crown 8vo, cloth **7/6**

MARINE ENGINEER'S POCKET-BOOK.

Consisting of useful Tables and Formulæ. By FRANK PROCTOR, A.I.N.A. Third Edition. Royal 32mo, leather, gilt edges, with strap . . . **4/0**

"We recommend it to our readers as going far to supply a long-felt want."—*Naval Science.*
"A most useful companion to all marine engineers."—*United Service Gazette.*

ELEMENTARY ENGINEERING.

A Manual for Young Marine Engineers and Apprentices. In the Form of Questions and Answers on Metals, Alloys, Strength of Materials, Construction and Management of Marine Engines and Boilers, Geometry, &c., &c. With an Appendix of Useful Tables. By J. S. BREWER. Crown 8vo, cloth . **1/6**

"Contains much valuable information for the class for whom it is intended, especially in the chapters on the management of boilers and engines."—*Nautical Magazine.*

PRACTICAL NAVIGATION.

Consisting of THE SAILOR'S SEA-BOOK, by JAMES GREENWOOD and W. H. ROSSER; together with the exquisite Mathematical and Nautical Tables for the Working of the Problems, by HENRY LAW, C.E., and Professor J. R. YOUNG. Illustrated. 12mo, strongly half-bound **7/0**

MARINE ENGINEER'S DRAWING-BOOK.

Adapted to the Requirements of the Board of Trade Examinations. By JOHN LOCKIE, C.E. With 22 Plates, Drawn to Scale. Royal 8vo, cloth . **3/6**

THE ART AND SCIENCE OF SAILMAKING.

By SAMUEL B. SADLER, Practical Sailmaker, late in the employment of Messrs. Ratsey and Lapthorne, of Cowes and Gosport. With Plates and other Illustrations. Small 4to, cloth **12/6**

"This extremely practical work gives a complete education in all the branches of the manufacture, cutting out, roping, seaming, and goring. It is copiously illustrated, and will form a first-rate text-book and guide."—*Portsmouth Times.*

CHAIN CABLES AND CHAINS.

Comprising Sizes and Curves of Links, Studs, &c., Iron for Cables and Chains, Chain Cable and Chain Making, Forming and Welding Links, Strength of Cables and Chains, Certificates for Cables, Marking Cables, Prices of Chain Cables and Chains, Historical Notes, Acts of Parliament, Statutory Tests, Charges for Testing, List of Manufacturers of Cables, &c., &c. By THOMAS W. TRAILL, F.E.R.N., M.Inst.C.E., Engineer-Surveyor-in-Chief, Board of Trade, Inspector of Chain Cable and Anchor Proving Establishments, and General Superintendent Lloyd's Committee on Proving Establishments. With numerous Tables, Illustrations, and Lithographic Drawings. Folio, cloth, bevelled boards . . . . **£2 2s.**

"It contains a vast amount of valuable information. Nothing seems to be wanting to make it complete and standard work of reference on the subject."—*Nautical Magazine.*

MINING AND METALLURGY.

COLLIERY WORKING AND MANAGEMENT.
Comprising the Duties of a Colliery Manager, the Oversight and Arrangement of Labour and Wages, and the different Systems of Working Coal Seams. By H. F. BULMAN and R. A. S. REDMAYNE. 350 pp., with 28 Plates and other Illustrations, including Underground Photographs. Medium 8vo, cloth. [*Just Published.* **15/0**

" This is, indeed, an admirable Handbook for Colliery Managers, in fact it is an indispensable adjunct to a Colliery Manager's education, as well as being a most useful and interesting work on the subject for all who in any way have to do with coal mining. The underground photographs are an attractive feature of the work, being very lifelike and necessarily true representations of the scenes they depict."—*Colliery Guardian.*

" Mr. Bulman and Mr. Redmayne, who are both experienced Colliery Managers of great literary ability, are to be congratulated on having supplied an authoritative work dealing with a side of the subject of coal mining which has hitherto received but scant treatment. The authors elucidate their text by 119 woodcuts and 28 plates, most of the latter being admirable reproductions of photographs taken underground with the aid of the magnesium flash-light. These illustrations are excellent."—*Nature.*

INFLAMMABLE GAS AND VAPOUR IN THE AIR
(The Detection and Measurement of). By FRANK CLOWES, D.Sc., Lond., F.I.C., Prof. of Chemistry in the University College, Nottingham. With a Chapter on THE DETECTION AND MEASUREMENT OF PETROLEUM VAPOUR by BOVERTON REDWOOD, F.R.S.E., Consulting Adviser to the Corporation of London under the Petroleum Acts. Crown 8vo, cloth.
[*Just Published.* Net. **5/0**

" Professor Clowes has given us a volume on a subject of much industrial importance . . . Those interested in these matters may be recommended to study this book, which is easy of comprehension and contains many good things."—*The Engineer.*

" A convenient summary of the work on which Professor Clowes has been engaged for some considerable time. . . . It is hardly necessary to say that any work on these subjects with these names on the title-page must be a valuable one, and one that no mining engineer—certainly no coal miner—can afford to ignore or to leave unread."—*Mining Journal.*

MACHINERY FOR METALLIFEROUS MINES.
A Practical Treatise for Mining Engineers, Metallurgists, and Managers of Mines. By E. HENRY DAVIES, M.E., F.G.S. Crown 8vo, 580 pp., with upwards of 300 Illustrations, cloth. [*Just Published.* **12/6**

" Mr. Davies, in this handsome volume, has done the advanced student and the manager of mines good service. Almost every kind of machinery in actual use is carefully described, and the woodcuts and plates are good."—*Athenæum.*

" From cover to cover the work exhibits all the same characteristics which excite the confidence and attract the attention of the student as he peruses the first page. The work may safely be recommended. By its publication the literature connected with the industry will be enriched and the reputation of its author enhanced."—*Mining Journal.*

METALLIFEROUS MINERALS AND MINING.
By D. C. DAVIES, F.G.S., Mining Engineer, &c., Author of "A Treatise on Slate and Slate Quarrying." Fifth Edition, thoroughly Revised and much Enlarged by his Son, E. HENRY DAVIES, M.E., F.G.S. With about 150 Illustrations. Crown 8vo, cloth **12/6**

" Neither the practical miner nor the general reader, interested in mines, can have a better book for his companion and his guide."—*Mining Journal.*

" We are doing our readers a service in calling their attention to this valuable work."—*Mining World.*

" As a history of the present state of mining throughout the world this book has a real value and it supplies an actual want."—*Athenæum.*

EARTHY AND OTHER MINERALS AND MINING.
By D. C. DAVIES, F.G.S., Author of " Metalliferous Minerals," &c. Third Edition, Revised and Enlarged by his Son, E. HENRY DAVIES, M.E., F.G.S. With about 100 Illustrations. Crown 8vo, cloth **12/6**

" We do not remember to have met with any English work on mining matters that contains the same amount of information packed in equally convenient form."—*Academy.*

" We should be inclined to rank it as among the very best of the handy technical and trades manuals which have recently appeared."—*British Quarterly Review.*

BRITISH MINING.

A Treatise on the History, Discovery, Practical Development, and Future Prospects of Metalliferous Mines in the United Kingdom. By ROBERT HUNT, F.R.S., late Keeper of Mining Records. Upwards of 950 pp., with 230 Illustrations. Second Edition, Revised. Super-royal 8vo, cloth **£2 2s.**

" The book is a treasure-house of statistical information on mining subjects, and we know of no other work embodying so great a mass of matter of this kind. Were this the only merit of Mr. Hunt's volume it would be sufficient to render it indispensable in the library of every one interested in the development of the mining and metallurgical industries of this country."—*Athenæum.*

" A mass of information not elsewhere available, and of the greatest value to those who may be interested in our great mineral industries."—*Engineer.*

MINE DRAINAGE.

A Complete and Practical Treatise on Direct-Acting Underground Steam Pumping Machinery, with a Description of a large number of the best known Engines, their General Utility and the Special Sphere of their Action, the Mode of their Application, and their merits compared with other forms of Pumping Machinery. By STEPHEN MICHELL. 8vo, cloth . **15/0**

" Will be highly esteemed by colliery owners and lessees, mining engineers, and students generally who require to be acquainted with the best means of securing the drainage of mines. It is a most valuable work, and stands almost alone in the literature of steam pumping machinery." —*Colliery Guardian.*

" Much valuable information is given, so that the book is thoroughly worthy of an extensive circulation amongst practical men and purchasers of machinery."—*Mining Journal.*

THE PROSPECTOR'S HANDBOOK.

A Guide for the Prospector and Traveller in search of Metal-Bearing or other Valuable Minerals. By J. W. ANDERSON, M.A. (Camb.), F.R.G.S., Author of " Fiji and New Caledonia." Seventh Edition, thoroughly Revised and much Enlarged. Small crown 8vo, cloth, **3/6** ; or, leather, pocket-book form, with tuck. *[Just Published.* **4/6.**

" Will supply a much-felt want, especially among Colonists, in whose way are so often thrown many mineralogical specimens the value of which it is difficult to determine."—*Engineer.*

" How to find commercial minerals, and how to identify them when they are found, are the leading points to which attention is directed. The author has managed to pack as much practical detail into his pages as would supply material for a book three times its size."—*Mining Journal.*

NOTES AND FORMULÆ FOR MINING STUDENTS.

By JOHN HERMAN MERIVALE, M.A., Late Professor of Mining in the Durham College of Science, Newcastle-upon-Tyne. Fourth Edition, Revised and Enlarged. By H. F. BULMAN, A.M.Inst.C.E. Small crown 8vo, cloth. *[Just Published.* **2/6**

" The author has done his work in a creditable manner, and has produced a book that will be of service to students and those who are practically engaged in mining operations."—*Engineer.*

THE MINER'S HANDBOOK.

A Handy Book of Reference on the subjects of Mineral Deposits, Mining Operations, Ore Dressing, &c. For the Use of Students and others interested in Mining Matters. By JOHN MILNE, F.R.S., Professor of Mining in the Imperial University of Japan. Revised Edition. Fcap. 8vo, leather . **7/6**

" Professor Milne's handbook is sure to be received with favour by all connected with mining, and will be extremely popular among students."—*Athenæum.*

POCKET-BOOK FOR MINERS AND METALLURGISTS.

Comprising Rules, Formulæ, Tables, and Notes for Use in Field and Office Work. By F. DANVERS POWER, F.G.S., M.E. Fcap. 8vo, leather . **9/0**

" This excellent book is an admirable example of its kind, and ought to find a large sale amongst English-speaking prospectors and mining engineers."—*Engineering.*

MINERAL SURVEYOR AND VALUER'S GUIDE.

Comprising a Treatise on Improved Mining Surveying and the Valuation of Mining Properties, with New Traverse Tables. By WM. LINTERN. Fourth Edition, Enlarged. 12mo, cloth **3/6**

THE COLLIERY MANAGER'S HANDBOOK.

A Comprehensive Treatise on the Laying-out and Working of Collieries, Designed as a Book of Reference for Colliery Managers, and for the Use of Coal Mining Students preparing for First-class Certificates. By CALEB PAMELY, Mining Engineer and Surveyor; Member of the North of England Institute of Mining and Mechanical Engineers; and Member of the South Wales Institute of Mining Engineers. With 700 Plans, Diagrams, and other Illustrations. Fourth Edition, Revised and Enlarged, medium 8vo, over 900 pp. Strongly bound . . . . **£1 5s.**

SUMMARY OF CONTENTS:—GEOLOGY.—SEARCH FOR COAL.—MINERAL LEASES AND OTHER HOLDINGS.—SHAFT SINKING.—FITTING UP THE SHAFT AND SURFACE ARRANGEMENTS.—STEAM BOILERS AND THEIR FITTINGS.—TIMBERING AND WALLING.—NARROW WORK AND METHODS OF WORKING. — UNDERGROUND CONVEYANCE. — DRAINAGE.—THE GASES MET WITH IN MINES; VENTILATION.—ON THE FRICTION OF AIR IN MINES.—THE PRIESTMAN OIL ENGINE; PETROLEUM AND NATURAL GAS.—SURVEYING AND PLANNING.—SAFETY LAMPS AND FIREDAMP DETECTORS.—SUNDRY AND INCIDENTAL OPERATIONS AND APPLIANCES.—COLLIERY EXPLOSIONS.—MISCELLANEOUS QUESTIONS AND ANSWERS.—*Appendix:* SUMMARY OF REPORT OF H.M. COMMISSIONERS ON ACCIDENTS IN MINES.

"Mr. Pamely has not only given us a comprehensive reference book of a very high order, suitable to the requirements of mining engineers and colliery managers, but has also provided mining students with a class-book that is as interesting as it is instructive."—*Colliery Manager.*

"Mr. Pamely's work is eminently suited to the purpose for which it is intended, being clear, interesting, exhaustive, rich in detail, and up to date, giving descriptions of the latest machines in every department. A mining engineer could scarcely go wrong who followed this work."—*Colliery Guardian.*

"This is the most complete 'all-round' work on coal-mining published in the English language. . . . No library of coal-mining books is complete without it."—*Colliery Engineer* (Scranton, Pa., U.S.A.).

COAL & IRON INDUSTRIES of the UNITED KINGDOM.

Comprising a Description of the Coal Fields, and of the Principal Seams of Coal, with Returns of their Produce and its Distribution, and Analyses of Special Varieties. Also, an Account of the Occurrence of Iron Ores in Veins or Seams; Analyses of each Variety; and a History of the Rise and Progress of Pig Iron Manufacture. By RICHARD MEADE. 8vo, cloth . . **£1 8s.**

"Of this book we may unreservedly say that it is the best of its class which we have ever met. . . . A book of reference which no one engaged in the iron or coal trades should omit from his library."—*Iron and Coal Trades Review.*

COAL AND COAL MINING.

By the late Sir WARINGTON W. SMYTH, M.A., F.R.S., &c., Chief Inspector of the Mines of the Crown. Seventh Edition, Revised and Enlarged. With numerous Illustrations, 12mo, cloth **3/6**

"As an outline is given of every known coal-field in this and other countries, as well as of the principal methods of working, the book will doubtless interest a very large number of readers."—*Mining Journal.*

ASBESTOS AND ASBESTIC.

Their Properties, Occurrence, and Use. By ROBERT H. JONES, F.S.A., Mineralogist, Hon. Mem. Asbestos Club, Black Lake, Canada. With Ten Collotype Plates and other Illustrations. Demy 8vo, cloth. [*Just Published.* **16/0**

"An interesting and invaluable work."—*Colliery Guardian.*

SUBTERRANEOUS SURVEYING

(Elementary and Practical Treatise on), with and without the Magnetic Needle. By THOMAS FENWICK, Surveyor of Mines, and THOMAS BAKER, C.E. Illustrated. 12mo, cloth **2/6**

GRANITES AND OUR GRANITE INDUSTRIES.

By GEORGE F. HARRIS, F.G.S., Membre de la Société Belge de Géologie, Lecturer on Economic Geology at the Birkbeck Institution, &c. With Illustrations. Crown 8vo, cloth **2/6**

"A clearly and well-written manual for persons engaged or interested in the granite industry."—*Scotsman.*

THE METALLURGY OF GOLD.

A Practical Treatise on the Metallurgical Treatment of Gold-bearing Ores. Including the Processes of Concentration, Chlorination, and Extraction by Cyanide, and the Assaying, Melting, and Refining of Gold. By M. EISSLER, Mining Engineer and Metallurgical Chemist, formerly Assistant Assayer of the U.S. Mint, San Francisco. Fourth Edition, Enlarged. With about 250 Illustrations and numerous Folding Plates and Working Drawings. Large crown 8vo, cloth. *[Just Published.* **16/0**

"This book thoroughly deserves its title of a 'Practical Treatise.' The whole process of gold milling, from the breaking of the quartz to the assay of the bullion, is described in clear and orderly narrative and with much, but not too much, fulness of detail."—*Saturday Review.*
"The work is a storehouse of information and valuable data, and we strongly recommend it to all professional men engaged in the gold-mining industry."—*Mining Journal.*

THE CYANIDE PROCESS OF GOLD EXTRACTION.

Including its Practical Application on the Witwatersrand Gold Fields in South Africa. By M. EISSLER, M.E., Author of "The Metallurgy of Gold," &c. With Diagrams and Working Drawings. Second Edition, Revised and Enlarged. 8vo, cloth. *[Just Published.* **7/6**

"This book is just what was needed to acquaint mining men with the actual working of a process which is not only the most popular, but is, as a general rule, the most successful for the extraction of gold from tailings."—*Mining Journal.*
"The work will prove invaluable to all interested in gold mining, whether metallurgists or as investors."—*Chemical News.*

THE METALLURGY OF SILVER.

A Practical Treatise on the Amalgamation, Roasting, and Lixiviation of Silver Ores. Including the Assaying, Melting, and Refining of Silver Bullion. By M. EISSLER, Author of "The Metallurgy of Gold," &c. Third Edition. Crown 8vo, cloth **10/6**

"A practical treatise, and a technical work which we are convinced will supply a long-felt want amongst practical men, and at the same time be of value to students and others indirectly connected with the industries."—*Mining Journal.*
"From first to last the book is thoroughly sound and reliable."—*Colliery Guardian.*
"For chemists, practical miners, assayers, and investors alike we do not know of any work on the subject so handy and yet so comprehensive."—*Glasgow Herald.*

THE METALLURGY OF ARGENTIFEROUS LEAD.

A Practical Treatise on the Smelting of Silver-Lead Ores and the Refining of Lead Bullion. Including Reports on various Smelting Establishments and Descriptions of Modern Smelting Furnaces and Plants in Europe and America. By M. EISSLER, M.E., Author of "The Metallurgy of Gold," &c. Crown 8vo, 400 pp., with 183 Illustrations, cloth **12/6**

"The numerous metallurgical processes, which are fully and extensively treated of, embrace all the stages experienced in the passage of the lead from the various natural states to its issue from the refinery as an article of commerce."—*Practical Engineer.*
"The present volume fully maintains the reputation of the author. Those who wish to obtain a thorough insight into the present state of this industry cannot do better than read this volume, and all mining engineers cannot fail to find many useful hints and suggestions in it."—*Industries.*

METALLURGY OF IRON.

By H. BAUERMAN, F.G.S., A.R.S.M. Sixth Edition, Revised and Enlarged. 12mo, cloth **5/0**

THE IRON ORES of GREAT BRITAIN and IRELAND.

Their Mode of Occurrence, Age and Origin, and the Methods of Searching for and Working Them. With a Notice of some of the Iron Ores of Spain. By J. D. KENDALL, F.G.S., Mining Engineer. Crown 8vo, cloth . . **16/0**

"The author has a thorough practical knowledge of his subject, and has supplemented a careful study of the available literature by unpublished information derived from his own observations. The result is a very useful volume, which cannot fail to be of value to all interested in the iron industry of the country."—*Industries.*

ELECTRICITY, ELECTRICAL ENGINEERING, &c.

SUBMARINE TELEGRAPHS.

Their History, Construction, and Working. Founded in part on WÜNSCHEN-DORFF'S "Traité de Télégraphie Sous—Marine," and Compiled from Authoritative and Exclusive Sources. By CHARLES BRIGHT, F.R.S.E. Super-royal 8vo, about 780 pp., fully Illustrated, including Maps and Folding Plates.
[*Just Published. Net.* **£3 3s.**

"There are few, if any, persons more fitted to write a treatise on submarine telegraphy than Mr. Charles Bright. The author has done his work admirably, and has written in a way which will appeal as much to the layman as to the engineer. This admirable volume must, for many years to come, hold the position of the English classic on submarine telegraphy."—*Engineer.*
"This book is full of Information. It makes a book of reference which should be in every engineer's library."—*Nature.*
"Mr. Bright's interestingly written and admirably illustrated book will meet with a welcome reception from cable men."—*Electrician.*
"The author deals with his subject from all points of view—political and strategical as well as scientific. The work will be of interest, not only to men of science, but to the general public. We can strongly recommend it."—*Athenæum.*
"The work contains a great store of technical information concerning the making and working of submarine telegraphs. In bringing together the most valuable results relating to the evolution of the telegraph, the author has rendered a service that will be very widely appreciated."—*Morning Post.*

THE ELECTRICAL ENGINEER'S POCKET-BOOK.

Consisting of Modern Rules, Formulæ, Tables, and Data. By H. R. KEMPE, M.Inst.E.E., A.M.Inst.C.E., Technical Officer Postal Telegraphs, Author of "A Handbook of Electrical Testing," "The Engineer's Year-Book," &c. Second Edition, thoroughly Revised, with Additions. With numerous Illustrations. Royal 32mo, oblong, leather **5/0**

"It is the best book of its kind."—*Electrical Engineer.*
"The Electrical Engineer's Pocket-Book is a good one."—*Electrician.*
"Strongly recommended to those engaged in the electrical industries."—*Electrical Review.*

ELECTRIC LIGHT FITTING.

A Handbook for Working Electrical Engineers, embodying Practical Notes on Installation Management. By J. W. URQUHART, Electrician, Author of "Electric Light," &c. With numerous Illustrations. Third Edition, Revised, with Additions. Crown 8vo, cloth. [*Just Published.* **5/0**

"This volume deals with what may be termed the mechanics of electric lighting, and is addressed to men who are already engaged in the work, or are training for it. The work traverses a great deal of ground, and may be read as a sequel to the same author's useful work on 'Electric Light.'"—*Electrician.*
"Eminently practical and useful. . . . Ought to be in the hands of every one in charge of an electric light plant."—*Electrical Engineer.*

ELECTRIC LIGHT.

Its Production and Use, Embodying Plain Directions for the Treatment of Dynamo-Electric Machines, Batteries, Accumulators, and Electric Lamps. By J. W. URQUHART, C.E. Sixth Edition, Revised, with Additions and 145 Illustrations. Crown 8vo, cloth. [*Just Published.* **7/6**

"The whole ground of electric lighting is more or less covered and explained in a very clear and concise manner."—*Electrical Review.*
"A *vade-mecum* of the salient facts connected with the science of electric lighting."—*Electrician.*
"You cannot for your purpose have a better book than 'Electric Light' by Urquhart."—*Engineer.*

DYNAMO CONSTRUCTION.

A Practical Handbook for the Use of Engineer-Constructors and Electricians-in-Charge. Embracing Framework Building, Field Magnet and Armature Winding and Grouping, Compounding, &c. With Examples of leading English, American, and Continental Dynamos and Motors. By J. W. URQUHART, Author of "Electric Light," &c. Second Edition, Enlarged. With 114 Illustrations. Crown 8vo, cloth **7/6**

"Mr. Urquhart's book is the first one which deals with these matters in such a way that the engineering student can understand them. The book is very readable, and the author leads his readers up to difficult subjects by reasonably simple tests."—*Engineering Review.*
"A book for which a demand has long existed."—*Mechanical World.*

THE MANAGEMENT OF DYNAMOS.

A Handy Book of Theory and Practice for the Use of Mechanics, Engineers, Students, and others in Charge of Dynamos. By G. W. LUMMIS PATERSON. With numerous Illustrations. Crown 8vo, cloth **3/6**

"An example which deserves to be taken as a model by other authors. The subject is treated in a manner which any intelligent man who is fit to be entrusted with charge of an engine should be able to understand. It is a useful book to all who make, tend, or employ electric machinery."—*Architect.*

THE STANDARD ELECTRICAL DICTIONARY.

A Popular Encyclopædia of Words and Terms Used in the Practice of Electrical Engineering. By T. O'CONOR SLOANE, A.M., Ph.D. Second Edition, with Appendix to date. Crown 8vo, 680 pp., 390 Illustrations, cloth.
[*Just Published.* **7/6**

"The work has many attractive features in it, and is, beyond doubt, a well put together and useful publication. The amount of ground covered may be gathered from the fact that in the index about 5,600 references will be found."—*Electrical Review.*

ELECTRIC SHIP-LIGHTING.

A Handbook on the Practical Fitting and Running of Ships' Electrical Plant. For the Use of Shipowners and Builders, Marine Electricians, and Seagoing Engineers-in-Charge. By J. W. URQUHART, C.E. With 88 Illustrations, Crown 8vo, cloth **7/6**

"The subject of ship electric lighting is one of vast importance, and Mr. Urquhart is to be highly complimented for placing such a valuable work at the service of marine electricians."—*The Steamship.*

ELECTRIC LIGHT FOR COUNTRY HOUSES.

A Practical Handbook on the Erection and Running of Small Installations, with Particulars of the Cost of Plant and Working. By J. H. KNIGHT. Second Edition, Revised. Crown 8vo, wrapper. [*Just Published.* **1/0**

"The book contains excellent advice and many practical hints for the help of those who wish to light their own houses."—*Building News.*

ELEMENTARY PRINCIPLES OF ELECTRIC LIGHTING.

By ALAN A. CAMPBELL SWINTON, Associate I.E.E. Third Edition, Enlarged and Revised. With 16 Illustrations. Crown 8vo, cloth . . . **1/6**

"Any one who desires a short and thoroughly clear exposition of the elementary principles of electric lighting cannot do better than read this little work."—*Bradford Observer.*

DYNAMIC ELECTRICITY AND MAGNETISM.

By PHILIP ATKINSON, A.M., Ph.D., Author of "Elements of Static Electricity," &c. Crown 8vo, 417 pp., with 120 Illustrations, cloth . **10/6**

THE ELECTRIC TRANSFORMATION OF POWER.

With its Application by the Electric Motor, including Electric Railway Construction. By P. ATKINSON, A.M., Ph.D. With 96 Illustrations. Crown 8vo, cloth **7/6**

HOW TO MAKE A DYNAMO.

A Practical Treatise for Amateurs. Containing numerous Illustrations and Detailed Instructions for Constructing a Small Dynamo to Produce the Electric Light. By ALFRED CROFTS. Fifth Edition, Revised and Enlarged. Crown 8vo, cloth. [*Just Published.* **2/0**

"The instructions given in this unpretentious little book are sufficiently clear and explicit to enable any amateur mechanic possessed of average skill and the usual tools to be found in an amateur's workshop to build a practical dynamo machine."—*Electrician.*

THE STUDENT'S TEXT-BOOK OF ELECTRICITY.

By H. M. NOAD, F.R.S. Cheaper Edition. 650 pp., with 470 Illustrations. Crown 8vo, cloth **9/0**

ARCHITECTURE, BUILDING, &c.

ARCHITECTURE, BUILDING, &c.

PRACTICAL BUILDING CONSTRUCTION.
A Handbook for Students Preparing for Examinations, and a Book of Reference for Persons Engaged in Building. By JOHN PARNELL ALLEN, Surveyor, Lecturer on Building Construction at the Durham College of Science, Newcastle-on-Tyne. Second Edition, Revised and Enlarged. Medium 8vo, 450 pp., with 1,000 Illustrations, cloth. [*Just Published.* **7/6**

"The most complete exposition of building construction we have seen. It contains all that is necessary to prepare students for the various examinations in building construction."—*Building News.*

"The author depends nearly as much on his diagrams as on his type. The pages suggest the hand of a man of experience in building operations—and the volume must be a blessing to many teachers as well as to students."—*The Architect.*

"The work is sure to prove a formidable rival to great and small competitors alike, and bids fair to take a permanent place as a favourite student's text-book. The large number of illustrations deserve particular mention for the great merit they possess for purposes of reference in exactly corresponding to convenient scales."—*Journal of the Royal Institute of British Architects.*

PRACTICAL MASONRY.
A Guide to the Art of Stone Cutting. Comprising the Construction, Setting Out, and Working of Stairs, Circular Work, Arches, Niches, Domes, Pendentives, Vaults, Tracery Windows, &c., &c. For the Use of Students, Masons, and other Workmen. By WILLIAM R. PURCHASE, Building Inspector to the Borough of Hove. Second Edition, with Glossary of Terms. Royal 8vo, 142 pp., with 52 Lithographic Plates, comprising nearly 400 separate Diagrams, cloth **7/6**

"Mr. Purchase's 'Practical Masonry' will undoubtedly be found useful to all interested in this important subject, whether theoretically or practically. Most of the examples given are from actual work carried out, the diagrams being carefully drawn. The book is a practical treatise on the subject, the author himself having commenced as an operative mason, and afterwards acted as foreman mason on many large and important buildings prior to the attainment of his present position. It should be found of general utility to architectural students and others, as well as to those to whom it is specially addressed."—*Journal of the Royal Institute of British Architects.*

"The author has evidently devoted much time and conscientious labour in the production of his book, which will be found very serviceable to students, masons, and other workmen, while its value is much enhanced by the capital illustrations, consisting of fifty lithographic plates, comprising about 400 diagrams."—*Illustrated Carpenter and Builder.*

CONCRETE: ITS NATURE AND USES.
A Book for Architects, Builders, Contractors, and Clerks of Works. By GEORGE L. SUTCLIFFE, A.R.I.B.A. 350 pp., with numerous Illustrations. Crown 8vo, cloth **7/6**

"The author treats a difficult subject in a lucid manner. The manual fills a long-felt gap. It is careful and exhaustive; equally useful as a student's guide and an architect's book of reference."—*Journal of the Royal Institute of British Architects.*

"There is room for this new book, which will probably be for some time the standard work on the subject for a builder's purpose."—*Glasgow Herald.*

THE MECHANICS OF ARCHITECTURE.
A Treatise on Applied Mechanics, especially Adapted to the Use of Architects. By E. W. TARN, M.A., Author of "The Science of Building," &c. Second Edition, Enlarged. Illustrated with 125 Diagrams. Crown 8vo, cloth **7/6**

"The book is a very useful and helpful manual of architectural mechanics, and really contains sufficient to enable a careful and painstaking student to grasp the principles bearing upon the majority of building problems. . . . Mr. Tarn has added, by this volume, to the debt of gratitude which is owing to him by architectural students for the many valuable works which he has produced for their use."—*The Builder.*

"The mechanics in the volume are really mechanics, and are harmoniously wrought in with the distinctive professional matter proper to the subject. The diagrams and type are commendably clear."—*The Schoolmaster.*

LOCKWOOD'S BUILDER'S PRICE BOOK for 1898.
A Comprehensive Handbook of the Latest Prices and Data for Builders, Architects, Engineers, and Contractors. Re-constructed, Re-written, and Greatly Enlarged. By FRANCIS T. W. MILLER. 800 closely-printed pages, crown 8vo, cloth **4/0**

"This book is a very useful one, and should find a place in every English office connected with the building and engineering professions."—*Industries.*

"An excellent book of reference."—*Architect.*

"In its new and revised form this Price Book is what a work of this kind should be—comprehensive, reliable, well arranged, legible, and well bound."—*British Architect.*

THE LONDON BUILDING ACT, 1894.

With the By-Laws and Regulations of the London County Council, and Introduction, Notes, Cases, and Index. By ALEX. J. DAVID, B.A., LL.M., of the Inner Temple, Barrister-at-Law. Crown 8vo, cloth **3/6**

"To all architects and district surveyors and builders Mr. David's manual will be welcome."
—*Building News.*

"The volume will doubtless be eagerly consulted by the building fraternity."—*Illustrated Carpenter and Builder.*

THE DECORATIVE PART OF CIVIL ARCHITECTURE.

By Sir WILLIAM CHAMBERS, F.R.S. With Portrait, Illustrations, Notes, and an EXAMINATION OF GRECIAN ARCHITECTURE, by JOSEPH GWILT, F.S.A. Revised and Edited by W. H. LEEDS. 66 Plates, 4to, cloth . . **21/0**

A HANDY BOOK OF VILLA ARCHITECTURE.

Being a Series of Designs for Villa Residences in various Styles. With Outline Specifications and Estimates. By C. WICKES, Architect, Author of "The Spires and Towers of England," &c. 61 Plates, 4to, half-morocco, gilt edges . . . . **£1 11s. 6d.**

"The whole of the designs bear evidence of their being the work of an artistic architect, and they will prove very valuable and suggestive."—*Building News.*

THE ARCHITECT'S GUIDE.

Being a Text-book of Useful Information for Architects, Engineers, Surveyors, Contractors, Clerks of Works, &c., &c. By FREDERICK ROGERS, Architect. Third Edition. Crown 8vo, cloth **3/6**

"As a text-book of useful information for architects, engineers, surveyors, &c., it would be hard to find a handier or more complete little volume."—*Standard.*

ARCHITECTURAL PERSPECTIVE.

The whole Course and Operations of the Draughtsman in Drawing a Large House in Linear Perspective. Illustrated by 43 Folding Plates. By F. O. FERGUSON. Second Edition, Enlarged. 8vo, boards . . . **3/6**

"It is the most intelligible of the treatises on this ill-treated subject that I have met with."—E. INGRESS BELL, ESQ., in the *R.I.B.A. Journal.*

PRACTICAL RULES ON DRAWING.

For the Operative Builder and Young Student in Architecture. By GEORGE PYNE. 14 Plates, 4to, boards **7/6**

MEASURING AND VALUING ARTIFICER'S WORK

(The Student's Guide to the Practice of). Containing Directions for taking Dimensions, Abstracting the same, and bringing the Quantities into Bill, with Tables of Constants for Valuation of Labour, and for the Calculation of Areas and Solidities. Originally edited by E. DOBSON, Architect. With Additions by E. W. TARN, M.A. Sixth Edition. With 8 Plates and 63 Woodcuts. Crown 8vo, cloth **7/6**

"This edition wil be found the most complete treatise on the principles of measuring and valuing artificer's work that has yet been published."—*Building News.*

TECHNICAL GUIDE, MEASURER, AND ESTIMATOR.

For Builders and Surveyors. Containing Technical Directions for Measuring Work in all the Building Trades, Complete Specifications for Houses, Roads, and Drains, and an Easy Method of Estimating the parts of a Building collectively. By A. C. BEATON. Eighth Edition. Waistcoat-pocket size, gilt edges **1/6**

"No builder, architect, surveyor, or valuer should be without his 'Beaton.'"—*Building News.*

CONSTRUCTIONAL IRON AND STEEL WORK.

As Applied to Public, Private, and Domestic Buildings. A Practical Treatise for Architects, Students, and Builders. By F. CAMPIN. Crown 8vo, cloth.
[*Just Published.* **3/6**

"Any one who wants a book on Ironwork, as employed in buildings for stanchions, columns, and beams, will find the present volume to be suitable. The author has had long and varied experience in designing this class of work. The illustrations have the character of working drawings. This practical book may be counted a most valuable work."—*British Architect.*

ARCHITECTURE, BUILDING, &c.

SPECIFICATIONS FOR PRACTICAL ARCHITECTURE.
A Guide to the Architect, Engineer, Surveyor, and Builder. With an Essay on the Structure and Science of Modern Buildings. Upon the Basis of the Work by ALFRED BARTHOLOMEW, thoroughly Revised, Corrected, and greatly added to by FREDERICK ROGERS, Architect. Third Edition, Revised. 8vo, cloth **15/0**
"The work is too well known to need any recommendation from us. It is one of the books with which every young architect must be equipped."—*Architect.*

THE SCIENCE OF BUILDING.
An Elementary Treatise on the Principles of Construction. By E. WYNDHAM TARN, M.A., Architect. Fourth Edition, with 59 Engravings. Fcap. 8vo, cloth **3/6**
"A very valuable book, which we strongly recommend to all students."—*Builder.*

THE HOUSE-OWNER'S ESTIMATOR.
Or, What will it Cost to Build, Alter, or Repair? A Price Book for Unprofessional People as well as the Architectural Surveyor and Builder. By J. D. SIMON. Edited by F. T. W. MILLER, A.R.I.B.A. Fourth Edition. Crown 8vo, cloth **3/6**
"In two years it will repay its cost a hundred times over."—*Field.*

A BOOK ON BUILDING.
Civil and Ecclesiastical, including Church Restoration; with the Theory of Domes and the Great Pyramid, &c. By Sir EDMUND BECKETT, Bart., LL.D., F.R.A.S. Second Edit. Fcap. 8vo, cloth **4/6**
"A book which is always amusing and nearly always instructive."—*The Times.*

SANITARY ARRANGEMENT OF DWELLING-HOUSES.
A Handbook for Householders and Owners of Houses. By A. J. WALLIS-TAYLER, A.M.Inst.C.E. With Illustrations. Crown 8vo, cloth . **2/6**
"This book will be largely read; it will be of considerable service to the public. It is well arranged, easily read, and for the most part devoid of technical terms."—*Lancet.*

VENTILATION.
A Text-book to the Practice of the Art of Ventilating Buildings. By W. P. BUCHAN, R.P. 12mo, cloth **3/6**
"Contains a great amount of useful practical information as thoroughly interesting as it is technically reliable."—*British Architect.*

PLUMBING.
A Text-book to the Practice of the Art or Craft of the Plumber. By W. P. BUCHAN, R.P. Seventh Edition, Enlarged. Crown 8vo, cloth . . **3/6**
"A text-book which may be safely put in the hands of every young plumber."—*Builder.*

PRACTICAL GEOMETRY.
For the Architect, Engineer, and Mechanic. Giving Rules for the Delineation and Application of various Geometrical Lines, Figures, and Curves. By E. W. TARN, M.A., Architect. 8vo, cloth **9/0**
"No book with the same objects in view has ever been published in which the clearness of the rules laid down and the illustrative diagrams have been so satisfactory."—*Scotsman.*

THE GEOMETRY OF COMPASSES.
Or, Problems Resolved by the mere Description of Circles and the Use of Coloured Diagrams and Symbols. By OLIVER BYRNE. Coloured Plates. Crown 8vo, cloth **3/6**

CARPENTRY, TIMBER, &c.

THE ELEMENTARY PRINCIPLES OF CARPENTRY.

A Treatise on the Pressure and Equilibrium of Timber Framing, the Resistance of Timber, and the Construction of Floors, Arches, Bridges, Roofs, Uniting Iron and Stone with Timber, &c. To which is added an Essay on the Nature and Properties of Timber, &c., with Descriptions of the kinds of Wood used in Building; also numerous Tables of the Scantlings of Timber for different purposes, the Specific Gravities of Materials, &c. By THOMAS TREDGOLD, C.E. With an Appendix of Specimens of Various Roofs of Iron and Stone, Illustrated. Seventh Edition, thoroughly Revised and considerably Enlarged by E. WYNDHAM TARN, M.A., Author of "The Science of Building," &c. With 61 Plates, Portrait of the Author, and several Woodcuts. In One large Vol., 4to, cloth **25/0**

"Ought to be in every architect's and every builder's library."—*Builder*.
"A work whose monumental excellence must commend it wherever skilful carpentry is concerned. The author's principles are rather confirmed than impaired by time. The additional plates are of great intrinsic value."—*Building News*.

WOODWORKING MACHINERY.

Its Rise, Progress, and Construction. With Hints on the Management of Saw Mills and the Economical Conversion of Timber. Illustrated with Examples of Recent Designs by leading English, French, and American Engineers. By M. POWIS BALE, A.M.Inst.C.E., M.I.M.E. Second Edition, Revised, with large Additions, large crown 8vo, 440 pp., cloth . . . **9/0**

"Mr. Bale is evidently an expert on the subject, and he has collected so much information that his book is all-sufficient for builders and others engaged in the conversion of timber."—*Architect*.
"The most comprehensive compendium of wood-working machinery we have seen. The author is a thorough master of his subject."—*Building News*.

SAW MILLS.

Their Arrangement and Management, and the Economical Conversion of Timber. By M. POWIS BALE, A.M.Inst.C.E. Second Edition, Revised. Crown 8vo, cloth. [*Just Published*. **10/6**

"The *administration* of a large sawing establishment is discussed, and the subject examined from a financial standpoint. Hence the size, shape, order, and disposition of saw mills and the like are gone into in detail, and the course of the timber is traced from its reception to its delivery in its converted state. We could not desire a more complete or practical treatise."—*Builder*.

THE CARPENTER'S NEW GUIDE.

Or, Book of Lines for Carpenters; comprising all the Elementary Principles essential for acquiring a knowledge of Carpentry. Founded on the late PETER NICHOLSON'S standard work. A New Edition, Revised by ARTHUR ASHPITEL, F.S.A. Together with Practical Rules on Drawing, by GEORGE PYNE. With 74 Plates, 4to, cloth . . . . **£1 1s.**

A PRACTICAL TREATISE ON HANDRAILING.

Showing New and Simple Methods for Finding the Pitch of the Plank, Drawing the Moulds, Bevelling, Jointing-up, and Squaring the Wreath. By GEORGE COLLINGS. Second Edition, Revised and Enlarged, to which is added A TREATISE ON STAIR-BUILDING. With Plates and Diagrams . . **2/6**

"Will be found of practical utility in the execution of this difficult branch of joinery."—*Builder*.
"Almost every difficult phase of this somewhat intricate branch of joinery is elucidated by the aid of plates and explanatory letterpress."—*Furniture Gazette*.

CIRCULAR WORK IN CARPENTRY AND JOINERY.

A Practical Treatise on Circular Work of Single and Double Curvature. By GEORGE COLLINGS. With Diagrams. Second Edition, 12mo, cloth . **2/6**

"An excellent example of what a book of this kind should be. Cheap in price, clear in definition, and practical in the examples selected."—*Builder*.

HANDRAILING COMPLETE IN EIGHT LESSONS.
On the Square-Cut System. By J. S. GOLDTHORP, Head of Building Department, Halifax Technical School. With Eight Plates and over 150 Practical Exercises. 4to, cloth **3/6**
"Likely to be of considerable value to joiners and others who take a pride in good work. The arrangement of the book is excellent. We heartily commend it to teachers and students."—*Timber Trades Journal.*

TIMBER MERCHANT'S and BUILDER'S COMPANION.
Containing New and Copious Tables of the Reduced Weight and Measurement of Deals and Battens, of all sizes, from One to a Thousand Pieces, and the relative Price that each size bears per Lineal Foot to any given Price per Petersburgh Standard Hundred ; the Price per Cube Foot of Square Timber to any given Price per Load of 50 Feet, &c., &c. By WILLIAM DOWSING. Fourth Edition, Revised and Corrected. Crown 8vo, cloth . . . **3/0**
"We are glad to see a fourth edition of these admirable tables, which for correctness and simplicity of arrangement leave nothing to be desired."—*Timber Trades Journal.*

THE PRACTICAL TIMBER MERCHANT.
A Guide for the Use of Building Contractors, Surveyors, Builders, &c., comprising useful Tables for all purposes connected with the Timber Trade, Marks of Wood, Essay on the Strength of Timber, Remarks on the Growth of Timber, &c. By W. RICHARDSON. Second Edition. Fcap. 8vo, cloth . **3/6**
"This handy manual contains much valuable information for the use of timber merchants, builders, foresters, and all others connected with the growth, sale, and manufacture of timber."—*Journal of Forestry.*

PACKING-CASE TABLES.
Showing the number of Superficial Feet in Boxes or Packing-Cases, from six inches square and upwards. By W. RICHARDSON, Timber Broker. Third Edition. Oblong 4to, cloth **3/6**
"Invaluable labour-saving tables."—*Ironmonger.*

GUIDE TO SUPERFICIAL MEASUREMENT.
Tables calculated from 1 to 200 inches in length by 1 to 108 inches in breadth. For the use of Architects, Surveyors, Engineers, Timber Merchants, Builders, &c. By JAMES HAWKINGS. Fourth Edition. Fcap., cloth . **3/6**
"A useful collection of tables to facilitate rapid calculation of surfaces. The exact area of any surface of which the limits have been ascertained can be instantly determined. The book will be found of the greatest utility to all engaged in building operations."—*Scotsman.*

PRACTICAL FORESTRY.
Its Bearing on the Improvement of Estates. By CHARLES E. CURTIS, F.S.I., F.S.S. Second Edition, Revised. Crown 8vo, cloth. [*Just Published.* **3/6**

THE ELEMENTS OF FORESTRY.
Designed to afford Information concerning the Planting and Care of Forest Trees for Ornament or Profit, with suggestions upon the Creation and Care of Woodlands. By F. B. HOUGH. Large crown 8vo, cloth . . . **10/0**

THE TIMBER IMPORTER'S, TIMBER MERCHANT'S, AND BUILDER'S STANDARD GUIDE.
By RICHARD E. GRANDY. Comprising:—An Analysis of Deal Standards, Home and Foreign, with Comparative Values and Tabular Arrangements for fixing Net Landed Cost on Baltic and North American Deals, including all intermediate Expenses, Freight, Insurance, &c. ; together with copious Information for the Retailer and Builder. Third Edition. 12mo, cloth . . **2/0**
"Everything it pretends to be : built up gradually, it leads one from a forest to a treenail and throws in as a makeweight a host of material concerning bricks, columns, cisterns, &c."—*English Mechanic.*

DECORATIVE ARTS, &c.

SCHOOL OF PAINTING FOR THE IMITATION OF WOODS AND MARBLES.
As Taught and Practised by A. R. VAN DER BURG and P. VAN DER BURG, Directors of the Rotterdam Painting Institution. Royal folio, 18½ by 12¼ in., Illustrated with 24 full-size Coloured Plates; also 12 plain Plates, comprising 154 Figures. Second and Cheaper Edition . . . . **£1 11s. 6d.**

LIST OF PLATES:—1. VARIOUS TOOLS REQUIRED FOR WOOD PAINTING.—2, 3. WALNUT; PRELIMINARY STAGES OF GRAINING AND FINISHED SPECIMEN.—4. TOOLS USED FOR MARBLE PAINTING AND METHOD OF MANIPULATION.—5, 6. ST. REMI MARBLE; EARLIER OPERATIONS AND FINISHED SPECIMEN. — 7. METHODS OF SKETCHING DIFFERENT GRAINS, KNOTS, &c.—8, 9. ASH; PRELIMINARY STAGES AND FINISHED SPECIMEN.—10. METHODS OF SKETCHING MARBLE GRAINS.—11, 12. BRECHE MARBLE; PRELIMINARY STAGES OF WORKING AND FINISHED SPECIMEN.—13. MAPLE; METHODS OF PRODUCING THE DIFFERENT GRAINS.—14, 15. BIRD'S-EYE MAPLE; PRELIMINARY STAGES AND FINISHED SPECIMEN.—16. METHODS OF SKETCHING THE DIFFERENT SPECIES OF WHITE MARBLE.—17, 18. WHITE MARBLE; PRELIMINARY STAGES OF PROCESS AND FINISHED SPECIMEN.—19. MAHOGANY; SPECIMENS OF VARIOUS GRAINS AND METHODS OF MANIPULATION.—20, 21. MAHOGANY; EARLIER STAGES AND FINISHED SPECIMEN.—22, 23, 24. SIENNA MARBLE; VARIETIES OF GRAIN, PRELIMINARY STAGES AND FINISHED SPECIMEN.—25, 26, 27. JUNIPER WOOD; METHODS OF PRODUCING GRAIN, &c.; PRELIMINARY STAGES AND FINISHED SPECIMEN.—28, 29, 30. VERT DE MER MARBLE; VARIETIES OF GRAIN AND METHODS OF WORKING, UNFINISHED AND FINISHED SPECIMENS.—31, 32, 33. OAK; VARIETIES OF GRAIN, TOOLS EMPLOYED AND METHODS OF MANIPULATION, PRELIMINARY STAGES AND FINISHED SPECIMEN.—34, 35, 36. WAULSORT MARBLE; VARIETIES OF GRAIN, UNFINISHED AND FINISHED SPECIMENS.

"Those who desire to attain skill in the art of painting woods and marbles will find advantage in consulting this book. . . . Some of the Working Men's Clubs should give their young men the opportunity to study it."—*Builder.*

"A comprehensive guide to the art. The explanations of the processes, the manipulation and management of the colours, and the beautifully executed plates will not be the least valuable to the student who aims at making his work a faithful transcript of nature."—*Building News.*

"Students and novices are fortunate who are able to become the possessors of so noble a work."—*The Architect.*

ELEMENTARY DECORATION.
A Guide to the Simpler Forms of Everyday Art. Together with PRACTICAL HOUSE DECORATION. By JAMES W. FACEY. With numerous Illustrations. In One Vol., strongly half-bound **5/0**

HOUSE PAINTING, GRAINING, MARBLING, AND SIGN WRITING.
A Practical Manual of. By ELLIS A. DAVIDSON. Seventh Edition. With Coloured Plates and Wood Engravings. 12mo, cloth boards . . . **6/0**

"A mass of information of use to the amateur and of value to the practical man."—*English Mechanic.*

THE DECORATOR'S ASSISTANT.
A Modern Guide for Decorative Artists and Amateurs, Painters, Writers, Gilders, &c. Containing upwards of 600 Receipts, Rules, and Instructions; with a variety of Information for General Work connected with every Class of Interior and Exterior Decorations, &c. Sixth Edition. 152 pp., cr. 8vo . **1/0**

"Full of receipts of value to decorators, painters, gilders, &c. The book contains the gist of larger treatises on colour and technical processes. It would be difficult to meet with a work so full of varied information on the painter's art."—*Building News.*

MARBLE DECORATION
And the Terminology of British and Foreign Marbles. A Handbook for Students. By GEORGE H. BLAGROVE, Author of "Shoring and its Application," &c. With 28 Illustrations. Crown 8vo, cloth **3/6**

"This most useful and much wanted handbook should be in the hands of every architect and builder."—*Building World.*

"A carefully and usefully written treatise; the work is essentially practical."—*Scotsman.*

DECORATIVE ARTS, &c. 31

DELAMOTTE'S WORKS ON ALPHABETS AND ILLUMINATION.

ORNAMENTAL ALPHABETS, ANCIENT & MEDIÆVAL.
From the Eighth Century, with Numerals; including Gothic, Church-Text, large and small, German, Italian, Arabesque, Initials for Illumination, Monograms, Crosses, &c., &c., for the use of Architectural and Engineering Draughtsmen, Missal Painters, Masons, Decorative Painters, Lithographers, Engravers, Carvers, &c., &c. Collected and Engraved by F. DELAMOTTE, and printed in Colours. New and Cheaper Edition. Royal 8vo, oblong, ornamental boards **2/6**
"For those who insert enamelled sentences round gilded chalices, who blazon shop legends over shop-doors, who letter church walls with pithy sentences from the Decalogue, this book will be useful."—*Athenæum*.

MODERN ALPHABETS, PLAIN AND ORNAMENTAL.
Including German, Old English, Saxon, Italic, Perspective, Greek, Hebrew, Court Hand, Engrossing, Tuscan, Riband, Gothic, Rustic, and Arabesque; with several Original Designs, and an Analysis of the Roman and Old English Alphabets, large and small, and Numerals, for the use of Draughtsmen, Surveyors, Masons, Decorative Painters, Lithographers, Engravers, Carvers, &c. Collected and Engraved by F. DELAMOTTE, and printed in Colours. New and Cheaper Edition. Royal 8vo, oblong, ornamental boards . **2/6**
"There is comprised in it every possible shape into which the letters of the alphabet and numerals can be formed, and the talent which has been expended in the conception of the various plain and ornamental letters is wonderful."—*Standard*.

MEDIÆVAL ALPHABETS AND INITIALS FOR ILLUMINATORS.
By F. G. DELAMOTTE. Containing 21 Plates and Illuminated Title, printed in Gold and Colours. With an Introduction by J. WILLIS BROOKS. Fourth and Cheaper Edition. Small 4to, ornamental boards **4/0**
"A volume in which the letters of the alphabet come forth glorified in gilding and all the colours of the prism interwoven and intertwined and intermingled."—*Sun*.

A PRIMER OF THE ART OF ILLUMINATION.
For the Use of Beginners; with a Rudimentary Treatise on the Art, Practical Directions for its Exercise, and Examples taken from Illuminated MSS., printed in Gold and Colours. By F. DELAMOTTE. New and Cheaper Edition. Small 4to, ornamental boards **6/0**
"The examples of ancient MSS. recommended to the student, which, with much good sense, the author chooses from collections accessible to all, are selected with judgment and knowledge as well as taste."—*Athenæum*.

THE EMBROIDERER'S BOOK OF DESIGN.
Containing Initials, Emblems, Cyphers, Monograms, Ornamental Borders, Ecclesiastical Devices, Mediæval and Modern Alphabets, and National Emblems. Collected by F. DELAMOTTE, and printed in Colours. Oblong royal 8vo, ornamental wrapper **1/6**
"The book will be of great assistance to ladies and young children who are endowed with the art of plying the needle in this most ornamental and useful pretty work."—*East Anglian Times*.

INSTRUCTIONS IN WOOD-CARVING FOR AMATEURS.
With Hints on Design. By A LADY. With 10 Plates. New and Cheaper Edition. Crown 8vo, in emblematic wrapper **2/0**
"The handicraft of the wood-carver, so well as a book can impart it, may be learnt from 'A Lady's' publication."—*Athenæum*.

PAINTING POPULARLY EXPLAINED.
By THOMAS JOHN GULLICK, Painter, and JOHN TIMBS, F.S.A. Including Fresco, Oil, Mosaic, Water-Colour, Water-Glass, Tempera, Encaustic, Miniature, Painting on Ivory, Vellum, Pottery, Enamel, Glass, &c. Fifth Edition. Crown 8vo, cloth **5/0**
*** *Adopted as a Prize Book at South Kensington.*

"Much may be learned, even by those who fancy they do not require to be taught, from the careful perusal of this unpretending but comprehensive treatise."—*Art Journal*.

NATURAL SCIENCE, &c.

THE VISIBLE UNIVERSE.

Chapters on the Origin and Construction of the Heavens. By J. E. GORE, F.R.A.S., Author of "Star Groups," &c. Illustrated by 6 Stellar Photographs and 12 Plates. Demy 8vo, cloth **16/0**

"A valuable and lucid summary of recent astronomical theory, rendered more valuable and attractive by a series of stellar photographs and other illustrations."—*The Times.*

"In presenting a clear and concise account of the present state of our knowledge Mr. Gore has made a valuable addition to the literature of the subject."—*Nature.*

"Mr. Gore's 'Visible Universe' is one of the finest works on astronomical science that have recently appeared in our language. In spirit and in method it is scientific from cover to cover, but the style is so clear and attractive that it will be as acceptable and as readable to those who make no scientific pretensions as to those who devote themselves specially to matters astronomical."—*Leeds Mercury.*

STAR GROUPS.

A Student's Guide to the Constellations. By J. ELLARD GORE, F.R.A.S., M.R.I.A., &c., Author of "The Visible Universe," "The Scenery of the Heavens," &c. With 30 Maps. Small 4to, cloth **5/0**

"The volume contains thirty maps showing stars of the sixth magnitude—the usual naked-eye limit—and each is accompanied by a brief commentary adapted to facilitate recognition and bring to notice objects of special interest. For the purpose of a preliminary survey of the 'midnight pomp' of the heavens nothing could be better than a set of delineations averaging scarcely twenty square inches in area and including nothing that cannot at once be identified."—*Saturday Review.*

AN ASTRONOMICAL GLOSSARY.

Or, Dictionary of Terms used in Astronomy. With Tables of Data and Lists of Remarkable and Interesting Celestial Objects. By J. ELLARD GORE F.R.A.S., Author of " The Visible Universe," &c. Small crown 8vo, cloth.
2/6

"A very useful little work for beginners in astronomy, and not to be despised by more advanced students."—*The Times.*

"A very handy book . . . the utility of which is much increased by it valuable tables of astronomical data."—*Athenæum.*

THE MICROSCOPE.

Its Construction and Management. Including Technique, Photo-micrography, and the Past and Future of the Microscope. By Dr. HENRI VAN HEURCK. Re-Edited and Augmented from the Fourth French Edition, and Translated by WYNNE E. BAXTER, F.G.S. 400 pp., with upwards of 250 Woodcuts, imp. 8vo, cloth **18/0**

"A translation of a well-known work, at once popular and comprehensive."—*Times.*
"The translation is as felicitous as it is accurate."—*Nature.*

PHOTO-MICROGRAPHY.

By Dr. H. VAN HEURCK. Extracted from the above Work. Royal 8vo, with Illustrations, sewed **1/0**

ASTRONOMY.

By the late Rev. ROBERT MAIN, M.A., F.R.S. Third Edition, Revised by WILLIAM THYNNE LYNN, B.A., F.R.A.S., formerly of the Royal Observatory, Greenwich. 12mo, cloth **2/0**

"A sound and simple treatise, very carefully edited, and a capital book for beginners."—*Knowledge.*

"Accurately brought down to the requirements of the present time by Mr. Lynn."—*Educational Times.*

A MANUAL OF THE MOLLUSCA.

A Treatise on Recent and Fossil Shells. By S. P. WOODWARD, A.L.S., F.G.S. With an Appendix on RECENT AND FOSSIL CONCHOLOGICAL DISCOVERIES, by RALPH TATE, A.L.S., F.G.S. With 23 Plates and upwards of 300 Woodcuts. Reprint of Fourth Edition (1880). Crown 8vo, cloth **7/6**

"A most valuable storehouse of conchological and geological information."—*Science Gossip.*

THE TWIN RECORDS OF CREATION.

Or, Geology and Genesis, their Perfect Harmony and Wonderful Concord. By G. W. V. LE VAUX. 8vo, cloth **5/0**

"A valuable contribution to the evidences of Revelation, and disposes very conclusively of the arguments of those who would set God's Works against God's Word. No real difficulty is shirked, and no sophistry is left unexposed."—*The Rock.*

NATURAL SCIENCE, &c.

HANDBOOK OF MECHANICS.
By Dr. LARDNER. Enlarged and re-written by BENJAMIN LOEWY, F.R.A.S. 378 Illustrations. Post 8vo, cloth **6/0**
"The perspicuity of the original has been retained, and chapters which had become obsolete have been replaced by others of more modern character. The explanations throughout are studiously popular, and care has been taken to show the application of the various branches of physics to the industrial arts, and to the practical business of life."—*Mining Journal.*

HANDBOOK OF HYDROSTATICS AND PNEUMATICS.
By Dr. LARDNER. New Edition, Revised and Enlarged by BENJAMIN LOEWY, F.R.A.S. With 236 Illustrations. Post 8vo, cloth **5/0**
"For those 'who desire to attain an accurate knowledge of physical science without the profound methods of mathematical investigation,' this work is well adapted."—*Chemical News.*

HANDBOOK OF HEAT.
By Dr. LARDNER. Edited and re-written by BENJAMIN LOEWY, F.R.A.S., &c. 117 Illustrations. Post 8vo, cloth **6/0**
"The style is always clear and precise, and conveys instruction without leaving any cloudiness or lurking doubts behind."—*Engineering.*

HANDBOOK OF OPTICS.
By Dr. LARDNER. New Edition. Edited by T. OLVER HARDING, B.A. Lond. With 298 Illustrations. Small 8vo, 448 pp., cloth **5/0**
"Written by one of the ablest English scientific writers, beautifully and elaborately illustrated."
—*Mechanics' Magazine.*

ELECTRICITY, MAGNETISM, AND ACOUSTICS.
By Dr. LARDNER. Edited by GEO. CAREY FOSTER, B.A., F.C.S. With 400 Illustrations. Small 8vo, cloth **5/0**
"The book could not have been entrusted to any one better calculated to preserve the terse and lucid style of Lardner, while correcting his errors and bringing up his work to the present state of scientific knowledge."—*Popular Science Review.*

HANDBOOK OF ASTRONOMY.
By Dr. LARDNER. Fourth Edition. Revised and Edited by EDWIN DUNKIN, F.R.A.S., Royal Observatory, Greenwich. With 38 Plates and upwards of 100 Woodcuts. 8vo, cloth **9/6**
"Probably no other book contains the same amount of information in so compendious and well arranged a form—certainly none at the price at which this is offered to the public."—*Athenæum.*
"We can do no other than pronounce this work a most valuable manual of astronomy, and we strongly recommend it to all who wish to acquire a general—but at the same time correct—acquaintance with this sublime science."—*Quarterly Journal of Science.*

MUSEUM OF SCIENCE AND ART.
Edited by Dr. LARDNER. With upwards of 1,200 Engravings on Wood. In Six Double Volumes, £1 1s. in a new and elegant cloth binding; or handsomely bound in half-morocco . . . . . . **£1 11s. 6d.**
"A cheap and interesting publication, alike informing and attractive. The papers combine subjects of importance and great scientific knowledge, considerable inductive powers, and a popular style of treatment."—*Spectator.*

Separate books formed from the above.

Common Things Explained. 5s. | Steam and its Uses. 2s. cloth.
The Microscope. 2s. cloth. | Popular Astronomy. 4s. 6d. cloth.
Popular Geology. 2s. 6d. cloth. | The Bee and White Ants. 2s. cloth.
Popular Physics. 2s. 6d. cloth. | The Electric Telegraph. 1s. 6d.

NATURAL PHILOSOPHY FOR SCHOOLS.
By Dr. LARDNER. Fcap. 8vo **3/6**
"A very convenient class book for Junior students in private schools."—*British Quarterly Review.*

ANIMAL PHYSIOLOGY FOR SCHOOLS.
By Dr. LARDNER. Fcap. 8vo **3/6**
"Clearly written, well arranged, and excellently illustrated."—*Gardener's Chronicle.*

THE ELECTRIC TELEGRAPH.
By Dr. LARDNER. Revised by E. B. BRIGHT, F.R.A.S. Fcap. 8vo. **2/6**
"One of the most readable books extant on the Electric Telegraph."—*English Mechanic.*

CHEMICAL MANUFACTURES, CHEMISTRY, &c.

THE GAS ENGINEER'S POCKET-BOOK.
Comprising Tables, Notes and Memoranda relating to the Manufacture, Distribution and Use of Coal Gas and the Construction of Gas Works. By H. O'CONNOR, A.M.Inst.C.E., 450 pp., crown 8vo, fully Illustrated, leather. [*Just Published.*] **10/6**

LIGHTING BY ACETYLENE
Generators, Burners, and Electric Furnaces. By WILLIAM E. GIBBS, M.E. With 66 Illustrations. Crown 8vo, cloth. [*Just Published.*] **7/6**

WATER AND ITS PURIFICATION.
A Handbook for the Use of Local Authorities, Sanitary Officers, and others interested in Water Supply. By S. RIDEAL, D.Sc. Lond., F.I.C. With numerous Illustrations and Tables. Crown 8vo, cloth. [*Just Published.*] **7/6**

"Dr. Rideal's book is both interesting and accurate, and contains a most useful *résumé* of the latest knowledge upon the subject of which it treats. It ought to be of great service to all who are connected with the supply of water for domestic or manufacturing purposes."—*The Engineer.*

"Dealing as clearly as it does with the various ramifications of such an important subject as water and its purification it may be warmly recommended. Local authorities and all engaged in sanitary affairs, and others interested in water supply, will read its pages with profit."—*Lancet.*

ENGINEERING CHEMISTRY.
A Practical Treatise for the Use of Analytical Chemists, Engineers, Iron Masters, Iron Founders, Students and others. Comprising Methods of Analysis and Valuation of the Principal Materials used in Engineering Work, with Analyses, Examples and Suggestions. By H. J. PHILLIPS, F.I.C., F.C.S. Second Edition, Enlarged. Crown 8vo, 400 pp., with Illustrations, cloth. **10/6**

"In this work the author has rendered no small service to a numerous body of practical men. . . . The analytical methods may be pronounced most satisfactory, being as accurate as the despatch required of engineering chemists permits."—*Chemical News.*

"Full of good things. As a handbook of technical analysis, it is very welcome."—*Builder.*

"The analytical methods given are, as a whole, such as are likely to give rapid and trustworthy results in experienced hands. . . . There is much excellent descriptive matter in the work, the chapter on 'Oils and Lubrication' being specially noticeable in this respect."—*Engineer.*

NITRO-EXPLOSIVES.
A Practical Treatise concerning the Properties, Manufacture, and Analysis of Nitrated Substances, including the Fulminates, Smokeless Powders, and Celluloid. By P. G. SANFORD, F.I.C., Consulting Chemist to the Cotton Powder Company, &c. With Illustrations. Crown 8vo, cloth. [*Just Published.*] **9/0**

"Any one having the requisite apparatus and materials could make nitro-glycerine or guncotton, to say nothing of other explosives, by the aid of the instructions in this volume. This is one of the very few text-books in which can be found just what is wanted. Mr. Sanford goes through the whole list of explosives commonly used, names any given explosive, and tells us of what it is composed and how it is manufactured. The book is excellent throughout."—*Engineer.*

A HANDBOOK ON MODERN EXPLOSIVES.
A Practical Treatise on the Manufacture and Use of Dynamite, Gun-Cotton, Nitro-Glycerine and other Explosive Compounds, including Collodion-Cotton. With Chapters on Explosives in Practical Application. By M. EISSLER, Mining Engineer and Metallurgical Chemist. Second Edition, Enlarged. With 150 Illustrations. Crown 8vo, cloth. [*Just Published.*] **12/6**

"Useful not only to the miner, but also to officers of both services to whom blasting and the use of explosives generally may at any time become a necessary auxiliary."—*Nature.*

DANGEROUS GOODS.
Their Sources and Properties, Modes of Storage and Transport. With Notes and Comments on Accidents arising therefrom, together with the Government and Railway Classifications, Acts of Parliament, &c. A Guide for the Use of Government and Railway Officials, Steamship Owners, Insurance Companies and Manufacturers, and Users of Explosives and Dangerous Goods. By H. JOSHUA PHILLIPS, F.I.C., F.C.S. Crown 8vo, 374 pp., cloth. **9/0**

"Merits a wide circulation, and an intelligent, appreciative study. —*Chemical News.*

CHEMICAL MANUFACTURES, CHEMISTRY, &c. 35

A MANUAL OF THE ALKALI TRADE.

Including the Manufacture of Sulphuric Acid, Sulphate of Soda, and Bleaching Powder. By JOHN LOMAS, Alkali Manufacturer, Newcastle-upon-Tyne and London. 390 pp. of Text. With 232 Illustrations and Working Drawings. Second Edition, with Additions. Super-royal 8vo, cloth . . **£1 10s.**

"This book is written by a manufacturer for manufacturers. The working details of the most approved forms of apparatus are given, and these are accompanied by no less than 232 wood engravings, all of which may be used for the purposes of construction. Every step in the manufacture is very fully described in this manual, and each improvement explained."—*Athenæum.*

"We find not merely a sound and luminous explanation of the chemical principles f the trade, but a notice of numerous matters which have a most important bearing on the successful conduct of alkali works, but which are greatly overlooked by even experienced technological authors."—*Chemical Review.*

THE BLOWPIPE IN CHEMISTRY, MINERALOGY, AND GEOLOGY.

Containing all known Methods of Anhydrous Analysis, many Working Examples, and Instructions for Making Apparatus. By Lieut.-Colonel W. A. ROSS, R.A., F.G.S. With 120 Illustrations. Second Edition, Enlarged. Crown 8vo, cloth **5/0**

"The student who goes conscientiously through the course of experimentation here laid down will gain a better insight into inorganic chemistry and mineralogy than if he had 'got up' any of the best text-books of the day, and passed any number of examinations in their contents."—*Chemical News.*

COMMERCIAL HANDBOOK OF CHEMICAL ANALYSIS.

Or, Practical Instructions for the Determination of the Intrinsic or Commercial Value of Substances used in Manufactures, in Trades, and in the Arts. By A. NORMANDY. New Edition by H. M. NOAD, Ph.D., F.R.S. Crown 8vo, cloth **12/6**

"We strongly recommend this book to our readers as a guide, alike indispensable to the housewife as to the pharmaceutical practitioner."—*Medical Times.*

THE MANUAL OF COLOURS AND DYE-WARES.

Their Properties, Applications, Valuations, Impurities and Sophistications. For the Use of Dyers, Printers, Drysalters, Brokers, &c. By J. W. SLATER. Second Edition, Revised and greatly Enlarged. Crown 8vo, cloth . **7/6**

"A complete encyclopædia of the *materia tinctoria.* The information given respecting each article is full and precise, and the methods of determining the value of articles such as these, so liable to sophistication, are given with clearness, and are practical as well as valuable."—*Chemist and Druggist.*

"There is no other work which covers precisely the same ground. To students preparing for examinations in dyeing and printing it will prove exceedingly useful."—*Chemical News.*

A HANDY BOOK FOR BREWERS.

Being a Practical Guide to the Art of Brewing and Malting. Embracing the Conclusions of Modern Research which bear upon the Practice of Brewing. By HERBERT EDWARDS WRIGHT, M.A. Second Edition, Enlarged. Crown 8vo, 530 pp., cloth. [*Just Published.* **12/6**

"May be consulted with advantage by the student who is preparing himself for examinational tests, while the scientific brewer will find in it a *résumé* of all the most important discoveries of modern times. The work is written throughout in a clear and concise manner, and the author takes great care to discriminate between vague theories and well-established facts."—*Brewers' Journal.*

"We have great pleasure in recommending this handy book, and have no hesitation in saying that it is one of the best—if not the best—which has yet been written on the subject of beer-brewing in this country; it should have a place on the shelves of every brewer's library."—*Brewers' Guardian.*

"Although the requirements of the student are primarily considered, an acquaintance of half-an-hour's duration cannot fail to impress the practical brewer with the sense of having found a trustworthy guide and practical counsellor in brewery matters."—*Chemical Trade Journal.*

FUELS: SOLID, LIQUID, AND GASEOUS.

Their Analysis and Valuation. For the Use of Chemists and Engineers. By H. J. PHILLIPS, F.C.S., formerly Analytical and Consulting Chemist to the G.E. Rlwy. Third Edition, Revised and Enlarged. Crown 8vo, cloth **2/0**

"Ought to have its place in the laboratory of every metallurgical establishment and wherever fuel is used on a large scale."—*Chemical News.*

C 2

36   CROSBY LOCKWOOD & SON'S CATALOGUE.

THE ARTISTS' MANUAL OF PIGMENTS.
Showing their Composition, Conditions of Permanency, Non-Permanency, and Adulterations; Effects in Combination with Each Other and with Vehicles; and the most Reliable Tests of Purity. By H. C. STANDAGE. Third Edition, crown 8vo, cloth **2/6**
"This work is indeed *multum-in-parvo*, and we can, with good conscience, recommend it to all who come in contact with pigments, whether as makers, dealers, or users."—*Chemical Review.*

A POCKET-BOOK OF MENSURATION AND GAUGING.
Containing Tables, Rules, and Memoranda for Revenue Officers, Brewers, Spirit Merchants, &c. By J. B. MANT, Inland Revenue. Second Edition, Revised. 18mo, leather **4/0**
"This handy and useful book is adapted to the requirements of the Inland Revenue Department, and will be a favourite book of reference. The range of subjects is comprehensive, and the arrangement simple and clear."—*Civilian.*
"Should be in the hands of every practical brewer."—*Brewers' Journal.*

INDUSTRIAL ARTS, TRADES, AND MANUFACTURES.

MODERN CYCLES.
A Practical Handbook on their Construction and Repair. By A. J. WALLIS-TAYLER, A. M. Inst. C. E. Author of "Refrigerating Machinery," &c. With upwards of 300 Illustrations. Crown 8vo, cloth.   [*Just Published.* **10/6**
"The large trade that is done in the component parts of bicycles has placed in the way of men mechanically inclined extraordinary facilities for building bicycles for their own use. . . . The book will prove a valuable guide for all those who aspire to the manufacture or repair of their own machines."—*The Field.*
"A most comprehensive and up-to-date treatise."—*The Cycle.*
"A very useful book, which is quite entitled to rank as a standard work for students of cycle construction."—*Wheeling.*

TEA PLANTING AND MANUFACTURE
(A Text Book of). Comprising Chapters on the History and Development of the Industry, the Cultivation of the Plant, the Preparation of the Leaf for the Market, the Botany and Chemistry of Tea, &c. With some Account of the Laws affecting Labour in Tea Gardens in Assam and elsewhere. By DAVID CROLE, late of the Jokai Tea Company, &c. With Plates and other Illustrations. Medium 8vo, cloth.   [*Just Published.* **16/0**
"The author writes as an expert, and gives the result of his personal experiences. The work can hardly fail to be of practical interest to tea growers and tea manufacturers."—*British Trade Journal.*

COTTON MANUFACTURE.
A Manual of Practical Instruction of the Processes of Opening, Carding, Combing, Drawing, Doubling and Spinning of Cotton, the Methods of Dyeing, &c. For the Use of Operatives, Overlookers, and Manufacturers. By JOHN LISTER, Technical Instructor, Pendleton. 8vo, cloth . . **7/6**
"This invaluable volume is a distinct advance in the literature of cotton manufacture."—*Machinery.*
"It is thoroughly reliable, fulfilling nearly all the requirements desired."—*Glasgow Herald.*

FLOUR MANUFACTURE.
A Treatise on Milling Science and Practice. By FRIEDRICH KICK, Imperial Regierungsrath, Professor of Mechanical Technology in the Imperial German Polytechnic Institute, Prague. Translated from the Second Enlarged and Revised Edition with Supplement. By H. H. P. POWLES, Assoc. Memb. Institution of Civil Engineers. Nearly 400 pp. Illustrated with 28 Folding Plates, and 167 Woodcuts. Royal 8vo, cloth . . . . . **£1 5s.**
"This valuable work is, and will remain, the standard authority on the science of milling. . . The miller who has read and digested this work will have laid the foundation, so to speak, of a successful career; he will have acquired a number of general principles which he can proceed to apply. In this handsome volume we at last have the accepted text-book of modern milling in good, sound English, which has little, if any, trace of the German idiom."—*The Miller.*
"The appearance of this celebrated work in English is very opportune, and British millers will, we are sure, no be slow in availing themselves of its pages."—*Millers' Gazette.*

INDUSTRIAL AND USEFUL ARTS. 37

CEMENTS, PASTES, GLUES, AND GUMS.
A Practical Guide to the Manufacture and Application of the various Agglutinants required in the Building, Metal-Working, Wood-Working, and Leather-Working Trades, and for Workshop, Laboratory or Office Use. With upwards of 900 Recipes and Formulæ. By H. C. STANDAGE, Chemist. Third Edition. Crown 8vo, cloth. [*Just Published.* **2/0**
"We have pleasure in speaking favourably of this volume. So far as we have had experience, which is not inconsiderable, this manual is trustworthy."—*Athenæum.*
"As a revelation of what are considered trade secrets, this book will arouse an amount of curiosity among the large number of industries it touches."—*Daily Chronicle.*

THE ART OF SOAP-MAKING.
A Practical Handbook of the Manufacture of Hard and Soft Soaps, Toilet Soaps, &c. Including many New Processes, and a Chapter on the Recovery of Glycerine from Waste Leys. By ALX. WATT. Fifth Edition, Revised, with an Appendix on Modern Candlemaking. Crown 8vo, cloth . . . **7/6**
"The work will prove very useful, not merely to the technological student, but to the practical soap boiler who wishes to understand the theory of his art."—*Chemical News.*
"A thoroughly practical treatise on an art which has almost no literature in our language. We congratulate the author on the success of his endeavour to fill a void in English technical literature."—*Nature.*

PRACTICAL PAPER-MAKING.
A Manual for Paper-Makers and Owners and Managers of Paper-Mills. With Tables, Calculations, &c. By G. CLAPPERTON, Paper-Maker. With Illustrations of Fibres from Micro-Photographs. Crown 8vo, cloth . . **5/0**
"The author caters for the requirements of responsible mill hands, apprentices, &c., whilst his manual will be found of great service to students of technology, as well as to veteran paper-makers and mill owners. The illustrations form an excellent feature."—*The World's Paper Trade Review.*
"We recommend everybody interested in the trade to get a copy of this thoroughly practical book."—*Paper Making.*

THE ART OF PAPER-MAKING.
A Practical Handbook of the Manufacture of Paper from Rags, Esparto, Straw, and other Fibrous Materials. Including the Manufacture of Pulp from Wood Fibre, with a Description of the Machinery and Appliances used. To which are added Details of Processes for Recovering Soda from Waste Liquors. By ALEXANDER WATT, Author of "The Art of Soap-Making." With Illustrations. Crown 8vo, cloth **7/6**
"It may be regarded as the standard work on the subject. The book is full of valuable nformation. The 'Art of Paper-Making' is in every respect a model of a text-book, either for a technical class, or for the private student."—*Paper and Printing Trades Journal.*

A TREATISE ON PAPER
For Printers and Stationers. With an Outline of Paper Manufacture; Complete Tables of Sizes, and Specimens of Different Kinds of Paper. By RICHARD PARKINSON, late of the Manchester Technical School. Demy 8vo, cloth.
[*Just Published.* **3/6**

THE ART OF LEATHER MANUFACTURE.
Being a Practical Handbook, in which the Operations of Tanning, Currying, and Leather Dressing are fully Described, and the Principles of Tanning Explained, and many Recent Processes Introduced; as also Methods for the Estimation of Tannin, and a Description of the Arts of Glue Boiling, Gut Dressing, &c. By ALEXANDER WATT, Author of "Soap-Making," &c. Fourth Edition. Crown 8vo, cloth **9/0**
"A sound, comprehensive treatise on tanning and its accessories. The book is an eminently valuable production, which redounds to the credit of both author and publishers."—*Chemical Review.*

THE ART OF BOOT AND SHOE MAKING.
A Practical Handbook, including Measurement, Last-Fitting, Cutting-Out, Closing and Making, with a Description of the most approved Machinery Employed. By JOHN B. LENO, late Editor of *St. Crispin*, and *The Boot and Shoe-Maker.* 12mo, cloth **2/0**

WOOD ENGRAVING.

A Practical and Easy Introduction to the Study of the Art. By W. N. BROWN. 12mo, cloth **1/6**

"The book is clear and complete, and will be useful to any one wanting to understand the first elements of the beautiful art of wood engraving."—*Graphic.*

MODERN HOROLOGY, IN THEORY AND PRACTICE.

Translated from the French of CLAUDIUS SAUNIER, ex-Director of the School of Horology at Macon, by JULIEN TRIPPLIN, F.R.A.S., Besancon Watch Manufacturer, and EDWARD RIGG, M.A., Assayer in the Royal Mint. With Seventy-eight Woodcuts and Twenty-two Coloured Copper Plates. Second Edition. Super-royal 8vo, cloth, **£2 2s.**; half-calf . . . **£2 10s.**

"There is no horological work in the English language at all to be compared to this production of M. Saunier's for clearness and completeness. It is alike good as a guide for the student and as a reference for the experienced horologist and skilled workman."—*Horological Journal.*

"The latest, the most complete, and the most reliable of those literary productions to which continental watchmakers are indebted for the mechanical superiority over their English brethren —in fact, the Book of Books, is M. Saunier's 'Treatise.'"—*Watchmaker, Jeweller, and Silversmith.*

THE WATCH ADJUSTER'S MANUAL.

A Practical Guide for the Watch and Chronometer Adjuster in Making, Springing, Timing and Adjusting for Isochronism, Positions and Temperatures. By C. E. FRITTS. 370 pp., with Illustrations, 8vo, cloth . . . **16/0**

THE WATCHMAKER'S HANDBOOK.

Intended as a Workshop Companion for those engaged in Watchmaking and the Allied Mechanical Arts. Translated from the French of CLAUDIUS SAUNIER, and enlarged by JULIEN TRIPPLIN, F.R.A.S., and EDWARD RIGG, M.A., Assayer in the Royal Mint. Third Edition. 8vo, cloth. **9/0**

"Each part is truly a treatise in itself. The arrangement is good and the language is clear and concise. It is an admirable guide for the young watchmaker."—*Engineering.*

"It is impossible to speak too highly of its excellence. It fulfils every requirement in a handbook intended for the use of a workman. Should be found in every workshop."—*Watch and Clockmaker.*

A HISTORY OF WATCHES & OTHER TIMEKEEPERS.

By JAMES F. KENDAL, M.B.H. Inst. Boards, **1/6**; or cloth, gilt . **2/6**

"The best which has yet appeared on this subject in the English language."—*Industries.*

"Open the book where you may, there is interesting matter in it concerning the ingenious devices of the ancient or modern horologer."—*Saturday Review.*

ELECTRO-DEPOSITION.

A Practical Treatise on the Electrolysis of Gold, Silver, Copper, Nickel, and other Metals and Alloys. With Descriptions of Voltaic Batteries, Magneto and Dynamo-Electric Machines, Thermopiles, and of the Materials and Processes used in every Department of the Art, and several Chapters on ELECTRO-METALLURGY. By ALEXANDER WATT, Author of "Electro-Metallurgy," &c. Third Edition, Revised. Crown 8vo, cloth . . **9/0**

"Eminently a book for the practical worker in electro-deposition. It contains practical descriptions of methods, processes and materials, as actually pursued and used in the workshop."—*Engineer.*

ELECTRO-METALLURGY.

Practically Treated. By ALEXANDER WATT. Tenth Edition, including the most recent Processes. 12mo, cloth **3/6**

"From this book both amateur and artisan may learn everything necessary for the successful prosecution of electroplating."—*Iron.*

JEWELLER'S ASSISTANT IN WORKING IN GOLD.

A Practical Treatise for Masters and Workmen, Compiled from the Experience of Thirty Years' Workshop Practice. By GEORGE E. GEE, Author of "The Goldsmith's Handbook," &c. Crown 8vo, cloth **7/6**

"This manual of technical education is apparently destined to be a valuable auxiliary to a handicraft which is certainly capable of great improvement."—*The Times.*

INDUSTRIAL AND USEFUL ARTS. 39

ELECTROPLATING.
A Practical Handbook on the Deposition of Copper, Silver, Nickel, Gold, Aluminium, Brass, Platinum, &c., &c. By J. W. URQUHART, C.E. Third Edition, Revised. Crown 8vo, cloth **5/0**
" An excellent practical manual."—*Engineering.*
" An excellent work, giving the newest information."—*Horological Journal.*

ELECTROTYPING.
The Reproduction and Multiplication of Printing Surfaces and Works of Art by the Electro-Deposition of Metals. By J. W. URQUHART, C.E. Crown 8vo, cloth **5/0**
" The book is thoroughly practical; the reader is, therefore, conducted through the leading laws of electricity, then through the metals used by electrotypers, the apparatus, and the depositing processes, up to the final preparation of the work."—*Art Journal.*

GOLDSMITH'S HANDBOOK.
By GEORGE E. GEE, Jeweller, &c. Fifth Edition. 12mo, cloth . . **3/0**
" A good, sound educator, and will be generally accepted as an authority."—*Horological Journal.*

SILVERSMITH'S HANDBOOK.
By GEORGE E. GEE, Jeweller, &c. Third Edition, with numerous Illustrations. 12mo, cloth **3/0**
" The chief merit of the work is its practical character. . . . The workers in the trade will speedily discover its merits when they sit down to study it."—*English Mechanic.*

*** *The above two works together, strongly half-bound, price 7s.*

SHEET METAL WORKER'S INSTRUCTOR.
Comprising a Selection of Geometrical Problems and Practical Rules for Describing the Various Patterns Required by Zinc, Sheet-Iron, Copper, and Tin-Plate Workers. By REUBEN HENRY WARN. New Edition, Revised and greatly Enlarged by JOSEPH G. HORNER, A.M.I.M.E. Crown 8vo, 254 pp., with 430 Illustrations, cloth. [*Just Published.*] **7/6**

BREAD & BISCUIT BAKER'S & SUGAR-BOILER'S ASSISTANT.
Including a large variety of Modern Recipes. With Remarks on the Art of Bread-making. By ROBERT WELLS. Third Edition. Crown 8vo, cloth . **2/0**
" A large number of wrinkles for the ordinary cook, as well as the baker."—*Saturday Review.*

PASTRYCOOK & CONFECTIONER'S GUIDE.
For Hotels, Restaurants, and the Trade in general, adapted also for Family Use. By R. WELLS, Author of " The Bread and Biscuit Baker." Crown 8vo, cloth **2/0**
" We cannot speak too highly of this really excellent work. In these days of keen competition our readers cannot do better than purchase this book."—*Bakers' Times.*

ORNAMENTAL CONFECTIONERY.
A Guide for Bakers, Confectioners and Pastrycooks; including a variety of Modern Recipes, and Remarks on Decorative and Coloured Work. With 129 Original Designs. By ROBERT WELLS. Second Edition. Crown 8vo . **5/0**
" A valuable work, practical, and should be in the hands of every baker and confectioner. The illustrative designs are alone worth treble the amount charged for the whole work."—*Bakers' Times.*

THE MODERN FLOUR CONFECTIONER, WHOLESALE AND RETAIL.
Containing a large Collection of Recipes for Cheap Cakes, Biscuits, &c. With remarks on the Ingredients Used in their Manufacture. By ROBERT WELLS, Author of " The Bread and Biscuit Baker," &c. Crown 8vo, cloth . **2/0**
" The work is of a decidedly practical character, and in every recipe regard is had to economical working."—*North British Daily Mail.*

RUBBER HAND STAMPS
And the Manipulation of Rubber. A Practical Treatise on the Manufacture of Indiarubber Hand Stamps, Small Articles of Indiarubber, The Hektograph, Special Inks, Cements, and Allied Subjects. By T. O'CONOR SLOANE, A.M., Ph.D. With numerous Illustrations. Square 8vo, cloth . . . **5/0**

HANDYBOOKS FOR HANDICRAFTS.
BY PAUL N. HASLUCK.
Editor of "Work" (New Series), Author of "Lathe Work," "Milling Machines," &c.
Crown 8vo, 144 pp., cloth, price 1s. each.

These HANDYBOOKS *have been written to supply information for* WORKMEN, STUDENTS, *and* AMATEURS *in the several Handicrafts, on the actual* PRACTICE *of the* WORKSHOP, *and are intended to convey in plain language* TECHNICAL KNOWLEDGE *of the several* CRAFTS. *In describing the processes employed, and the manipulation of material, workshop terms are used ; workshop practice is fully explained ; and the text is freely illustrated with drawings of modern tools, appliances, and processes.*

THE METAL TURNER'S HANDYBOOK.
A Practical Manual for Workers at the Foot-Lathe. With over 100 Illustrations. 1/0
"The book will be of service alike to the amateur and the artisan turner. It displays thorough knowledge of the subject."—*Scotsman.*

THE WOOD TURNER'S HANDYBOOK.
A Practical Manual for Workers at the Lathe. With over 100 Illustrations 1/0
"We recommend the book to young turners and amateurs. A multitude of workmen have hitherto sought in vain for a manual of this special industry."—*Mechanical World.*

THE WATCH JOBBER'S HANDYBOOK.
A Practical Manual on Cleaning, Repairing, and Adjusting. With upwards of 100 Illustrations 1/0
"We strongly advise all young persons connected with the watch trade to acquire and study is inexpensive work."—*Clerkenwell Chronicle.*

THE PATTERN MAKER'S HANDYBOOK.
A Practical Manual on the Construction of Patterns for Founders. With upwards of 100 Illustrations 1/0
"A most valuable, if not indispensable manual for the pattern maker."—*Knowledge.*

THE MECHANIC'S WORKSHOP HANDYBOOK.
A Practical Manual on Mechanical Manipulation, embracing Information on various Handicraft Processes. With Useful Notes and Miscellaneous Memoranda. Comprising about 200 Subjects 1/0
"A very clever and useful book, which should be found in every workshop; and it should certainly find a place in all technical schools."—*Saturday Review.*

THE MODEL ENGINEER'S HANDYBOOK.
A Practical Manual on the Construction of Model Steam Engines. With upwards of 100 Illustrations. 1/0
"Mr. Hasluck has produced a very good little book."—*Builder.*

THE CLOCK JOBBER'S HANDYBOOK.
A Practical Manual on Cleaning, Repairing, and Adjusting. With upwards of 100 Illustrations 1/0
"It is of inestimable service to those commencing the trade."—*Coventry Standard.*

THE CABINET MAKER'S HANDYBOOK.
A Practical Manual on the Tools, Materials, Appliances, and Processes employed in Cabinet Work. With upwards of 100 Illustrations . . 1/0
"Mr. Hasluck's thorough-going little Handybook is amongst the most practical guides we have seen for beginners in cabinet-work."—*Saturday Review.*

THE WOODWORKER'S HANDYBOOK OF MANUAL INSTRUCTION.
Embracing Information on the Tools, Materials, Appliances and Processes Employed in Woodworking. With 104 Illustrations. . . . 1/0

OPINIONS OF THE PRESS.
"Written by a man who knows, not only how work ought to be done, but how to do it, and how to convey his knowledge to others."—*Engineering.*
"Mr. Hasluck writes admirably, and gives complete instructions."—*Engineer.*
"Mr. Hasluck combines the experience of a practical teacher with the manipulative skill and scientific knowledge of processes of the trained mechanician, and the manuals are marvels of what can be produced at a popular price."—*Schoolmaster.*
"Helpful to workmen of all ages and degrees of experience."—*Daily Chronicle.*
"Practical, sensible, and remarkably cheap."—*Journal of Education.*
"Concise, clear and practical."—*Saturday Review.*

COMMERCE, COUNTING-HOUSE WORK, TABLES, &c.

LESSONS IN COMMERCE.
By Professor R. GAMBARO, of the Royal High Commercial School at Genoa. Edited and Revised by JAMES GAULT, Professor of Commerce and Commercial Law in King's College, London. Second Edition, Revised. Crown 8vo . **3/6**
"The publishers of this work have rendered considerable service to the cause of commercial education by the opportune production of this volume. . . . The work is peculiarly acceptable to English readers and an admirable addition to existing class books. In a phrase, we think the work attains its object in furnishing a brief account of those laws and customs of British trade with which the commercial man interested therein should be familiar."—*Chamber of Commerce Journal.*
"An invaluable guide in the hands of those who are preparing for a commercial career, and, in fact, the information it contains on matters of business should be impressed on every one."—*Counting House.*

THE FOREIGN COMMERCIAL CORRESPONDENT.
Being Aids to Commercial Correspondence in Five Languages—English, French, German, Italian, and Spanish. By CONRAD E. BAKER. Second Edition. Crown 8vo, cloth **3/6**
"Whoever wishes to correspond in all the languages mentioned by Mr. Baker cannot do better than study this work, the materials of which are excellent and conveniently arranged. They consist not of entire specimen letters, but—what are far more useful—short passages, sentences, or phrases expressing the same general idea in various forms."—*Athenæum.*
"A careful examination has convinced us that it is unusually complete, well arranged and reliable. The book is a thoroughly good one."—*Schoolmaster.*

A NEW BOOK OF COMMERCIAL FRENCH.
Grammar — Vocabulary — Correspondence — Commercial Documents — Geography—Arithmetic—Lexicon. By P. CARROUÉ, Professor in the City High School J.—B. Say (Paris). Crown 8vo, cloth **4/6**
"M. Carroué's book is a *vade mecum* of commercial French, and would be distinctly in its place in every merchant's office. Nothing better could be desired."—*Educational Times.*

FACTORY ACCOUNTS: their PRINCIPLES & PRACTICE.
A Handbook for Accountants and Manufacturers, with Appendices on the Nomenclature of Machine Details; the Income Tax Acts; the Rating of Factories; Fire and Boiler Insurance; the Factory and Workshop Acts, &c., including also a Glossary of Terms and a large number of Specimen Rulings. By EMILE GARCKE and J. M. FELLS. Fourth Edition, Revised and Enlarged. Demy 8vo, 250 pp., strongly bound **6/0**
"A very interesting description of the requirements of Factory Accounts. . . . The principle of assimilating the Factory Accounts to the general commercial books is one which we thoroughly agree with."—*Accountants' Journal.*
"Characterised by extreme thoroughness. There are few owners of factories who would not derive great benefit from the perusal of this most admirable work."—*Local Government Chronicle.*

MODERN METROLOGY.
A Manual of the Metrical Units and Systems of the present Century. With an Appendix containing a proposed English System. By LOWIS D. A. JACKSON, A. M. Inst. C. E., Author of "Aid to Survey Practice," &c. Large crown 8vo, cloth **12/6**
"We recommend the work to all interested in the practical reform of our weights and measures."—*Nature.*

A SERIES OF METRIC TABLES.
In which the British Standard Measures and Weights are compared with those of the Metric System at present in Use on the Continent. By C. H. DOWLING, C.E. 8vo, strongly bound **10/6**
"Mr. Dowling's Tables are well put together as a ready reckoner for the conversion of one system into the other."—*Athenæum.*

THE IRON AND METAL TRADES' COMPANION.
For Expeditiously Ascertaining the Value of any Goods bought or sold by Weight, from 1s. per cwt. to 112s. per cwt., and from one farthing per pound to one shilling per pound. By THOMAS DOWNIE. 396 pp., leather . . **9/0**
"A most useful set of tables, nothing like them before existed."—*Building News.*
"Although specially adapted to the iron and metal trades, the tables will be found useful every other business in which merchandise is bought and sold by weight."—*Railway News.*

NUMBER, WEIGHT, AND FRACTIONAL CALCULATOR.

Containing upwards of 250,000 Separate Calculations, showing at a Glance the Value at 422 Different Rates, ranging from 1/16th of a Penny to 20s. each, or per cwt., and £20 per ton, of any number of articles consecutively, from 1 to 470. Any number of cwts., qrs., and lbs., from 1 cwt. to 470 cwts. Any number of tons, cwts., qrs., and lbs., from 1 to 1,000 tons. By WILLIAM CHADWICK, Public Accountant. Third Edition, Revised. 8vo, strongly bound . **18/0**

"It is as easy of reference for any answer or any number of answers as a dictionary. For making up accounts or estimates the book must prove invaluable to all who have any considerable quantity of calculations involving price and measure in any combination to do."—*Engineer.*
"The most perfect work of the kind yet prepared."—*Glasgow Herald.*

THE WEIGHT CALCULATOR.

Being a Series of Tables upon a New and Comprehensive Plan, exhibiting at one Reference the exact Value of any Weight from 1 lb. to 15 tons, at 300 Progressive Rates, from 1d. to 168s. per cwt., and containing 186,000 Direct Answers, which, with their Combinations, consisting of a single addition mostly to be performed at sight), will afford an aggregate of 10,266,000 Answers; the whole being calculated and designed to ensure correctness and promote despatch. By HENRY HARBEN, Accountant. Fifth Edition, carefully Corrected. Royal 8vo, strongly half-bound . . . . . **£1 5s.**

"A practical and useful work of reference for men of business generally."—*Ironmonger.*
"Of priceless value to business men. It is a necessary book in all mercantile offices."—*Sheffield Independent.*

THE DISCOUNT GUIDE.

Comprising several Series of Tables for the Use of Merchants, Manufacturers, Ironmongers, and Others, by which may be ascertained the Exact Profit arising from any mode of using Discounts, either in the Purchase or Sale of Goods, and the method of either Altering a Rate of Discount, or Advancing a Price, so as to produce, by one operation, a sum that will realise any required Profit after allowing one or more Discounts: to which are added Tables of Profit or Advance from 1¼ to 90 per cent., Tables of Discount from 1¼ to 98¾ per cent., and Tables of Commission, &c., from ⅛ to 10 per cent. By HENRY HARBEN, Accountant. New Edition, Corrected. Demy 8vo, half-bound . **£1 5s.**

"A book such as this can only be appreciated by business men, to whom the saving of time means saving of money. The work must prove of great value to merchants, manufacturers, and general traders."—*British Trade Journal.*

TABLES OF WAGES.

At 54, 52, 50 and 48 Hours per Week. Showing the Amounts of Wages from One quarter of an hour to Sixty-four hours, in each case at Rates of Wages advancing by One Shilling from 4s. to 55s. per week. By THOS. GARBUTT, Accountant. Square crown 8vo, half-bound **6/0**

IRON-PLATE WEIGHT TABLES.

For Iron Shipbuilders, Engineers, and Iron Merchants. Containing the Calculated Weights of upwards of 150,000 different sizes of Iron Plates from 1 foot by 6 in. by ¼ in. to 10 feet by 5 feet by 1 in. Worked out on the Basis of 40 lbs. to the square foot of Iron of 1 inch in thickness. By H. BURLINSON and W. H. SIMPSON. 4to, half-bound . . . . . . **£1 5s.**

MATHEMATICAL TABLES (ACTUARIAL).

Comprising Commutation and Conversion Tables, Logarithms, Cologarithms, Antilogarithms and Reciprocals. By J. W. GORDON. Royal 8vo, mounted on canvas, in cloth case. [*Just Published.* **5/0**

AGRICULTURE, FARMING, GARDENING, &c.

THE COMPLETE GRAZIER AND FARMER'S AND CATTLE BREEDER'S ASSISTANT.

A Compendium of Husbandry. Originally Written by WILLIAM YOUATT, Thirteenth Edition, entirely Re-written, considerably Enlarged, and brought up to the Present Requirements of Agricultural Practice, by WILLIAM FREAM, LL.D., Steven Lecturer in the University of Edinburgh, Author of "The Elements of Agriculture," &c. Royal 8vo, 1,100 pp., with over 450 Illustrations, strongly and handsomely bound . . . **£1 11s. 6d.**

SUMMARY OF CONTENTS.

BOOK I. ON THE VARIETIES, BREEDING, REARING, FATTENING AND MANAGEMENT OF CATTLE.
BOOK II. ON THE ECONOMY AND MANAGEMENT OF THE DAIRY.
BOOK III. ON THE BREEDING, REARING, AND MANAGEMENT OF HORSES.
BOOK IV. ON THE BREEDING, REARING, AND FATTENING OF SHEEP.
BOOK V. ON THE BREEDING, REARING, AND FATTENING OF SWINE.
BOOK VI. ON THE DISEASES OF LIVE STOCK.
BOOK VII. ON THE BREEDING, REARING, AND MANAGEMENT OF POULTRY.
BOOK VIII. ON FARM OFFICES AND IMPLEMENTS OF HUSBANDRY.
BOOK IX. ON THE CULTURE AND MANAGEMENT OF GRASS LANDS.
BOOK X. ON THE CULTIVATION AND APPLICATION OF GRASSES, PULSE AND ROOTS.
BOOK XI. ON MANURES AND THEIR APPLICATION TO GRASS LAND AND CROPS.
BOOK XII. MONTHLY CALENDARS OF FARMWORK.

*** OPINIONS OF THE PRESS ON THE NEW EDITION.

"Dr. Fream is to be congratulated on the successful attempt he has made to give us a work which will at once become the standard classic of the farm practice of the country. We believe that it will be found that it has no compeer among the many works at present in existence. . . . The illustrations are admirable, while the frontispiece, which represents the well-known bull, New Year's Gift, bred by the Queen, is a work of art."—*The Times.*

"The book must be recognised as occupying the proud position of the most exhaustive work of reference in the English language on the subject with which it deals."—*Athenæum.*

"The most comprehensive guide to modern farm practice that exists in the English language to-day. . . . The book is one that ought to be on every farm and in the library of every land owner."—*Mark Lane Express.*

"In point of exhaustiveness and accuracy the work will certainly hold a pre-eminent and unique position among books dealing with scientific agricultural practice. It is, in fact, an agricultural library of itself."—*North British Agriculturist.*

"A compendium of authoritative and well-ordered knowledge on every conceivable branch of the work of the live stock farmer; probably without an equal in this or any other country."—*Yorkshire Post.*

FARM LIVE STOCK OF GREAT BRITAIN.

By ROBERT WALLACE, F.L.S., F.R.S.E., &c., Professor of Agriculture and Rural Economy in the University of Edinburgh. Third Edition, thoroughly Revised and considerably Enlarged. With over 120 Phototypes of Prize Stock. Demy 8vo, 384 pp., with 79 Plates and Maps, cloth. . . **12/6**

"A really complete work on the history, breeds, and management of the farm stock of Great Britain, and one which is likely to find its way to the shelves of every country gentleman's library."—*The Times.*

"The latest edition of 'Farm Live Stock of Great Britain' is a production to be proud of, and its issue not the least of the services which its author has rendered to agricultural science."—*Scottish Farmer.*

"The book is very attractive, . . . and we can scarcely imagine the existence of a farmer who would not like to have a copy of this beautiful and useful work."—*Mark Lane Express.*

NOTE-BOOK OF AGRICULTURAL FACTS & FIGURES FOR FARMERS AND FARM STUDENTS.

By PRIMROSE MCCONNELL, B.Sc., Fellow of the Highland and Agricultural Society, Author of "Elements of Farming." Sixth Edition, Re-written, Revised, and greatly Enlarged. Fcap. 8vo, 480 pp., leather. [*Just Published.* **6/0**

SUMMARY OF CONTENTS : SURVEYING AND LEVELLING.—WEIGHTS AND MEASURES.—MACHINERY AND BUILDINGS.—LABOUR.—OPERATIONS.—DRAINING.—EMBANKING.—GEOLOGICAL MEMORANDA.—SOILS.—MANURES.—CROPPING.—CROPS.—ROTATIONS.—WEEDS.—FEEDING.—DAIRYING.—LIVE STOCK.—HORSES.—CATTLE.—SHEEP.—PIGS.—POULTRY.—FORESTRY.—HORTICULTURE.—MISCELLANEOUS.

"No farmer, and certainly no agricultural student, ought to be without this *multum-in-parvo* manual of all subjects connected with the farm."—*North British Agriculturist.*

"This little pocket-book contains a large amount of useful information upon all kinds of agricultural subjects. Something of the kind has long been wanted."—*Mark Lane Express.*

"The amount of information it contains is most surprising ; the arrangement of the matter is so methodical—although so compressed—as to be intelligible to everyone who takes a glance through its pages. They teem with information."—*Farm and Home.*

BRITISH DAIRYING.

A Handy Volume on the Work of the Dairy-Farm. For the Use of Technical Instruction Classes, Students in Agricultural Colleges and the Working Dairy-Farmer. By Prof. J. P. SHELDON. With Illustrations. Second Edition, Revised. Crown 8vo, cloth. [*Just Published.* **2/6**

"Confidently recommended as a useful text-book on dairy farming."—*Agricultural Gazette.*
"Probably the best half-crown manual on dairy work that has yet been produced."—*North British Agriculturist.*
"It is the soundest little work we have yet seen on the subject."—*The Times.*

MILK, CHEESE, AND BUTTER.

A Practical Handbook on their Properties and the Processes of their Production. Including a Chapter on Cream and the Methods of its Separation from Milk. By JOHN OLIVER, late Principal of the Western Dairy Institute, Berkeley. With Coloured Plates and 200 Illustrations. Crown 8vo, cloth. **7/6**

"An exhaustive and masterly production. It may be cordially recommended to all students and practitioners of dairy science."—*North British Agriculturist.*
"We recommend this very comprehensive and carefully-written book to dairy-farmers and students of dairying. It is a distinct acquisition to the library of the agriculturist."—*Agricultural Gazette.*

SYSTEMATIC SMALL FARMING.

Or, The Lessons of My Farm. Being an Introduction to Modern Farm Practice for Small Farmers. By R. SCOTT BURN, Author of "Outlines of Modern Farming," &c. Crown 8vo, cloth. **6/0**

"This is the completest book of its class we have seen, and one which every amateur farmer will read with pleasure, and accept as a guide."—*Field.*

OUTLINES OF MODERN FARMING.

By R. SCOTT BURN. Soils, Manures, and Crops—Farming and Farming Economy—Cattle, Sheep, and Horses—Management of Dairy, Pigs, and Poultry—Utilisation of Town-Sewage, Irrigation, &c. Sixth Edition. In One Vol., 1,250 pp., half-bound, profusely Illustrated **12/0**

FARM ENGINEERING, The COMPLETE TEXT-BOOK of.

Comprising Draining and Embanking; Irrigation and Water Supply; Farm Roads, Fences and Gates; Farm Buildings; Barn Implements and Machines; Field Implements and Machines; Agricultural Surveying, &c. By Professor JOHN SCOTT. In One Vol., 1,150 pp., half-bound, with over 600 Illustrations. **12/0**

"Written with great care, as well as with knowledge and ability. The author has done his work well; we have found him a very trustworthy guide wherever we have tested his statements. The volume will be of great value to agricultural students."—*Mark Lane Express.*

THE FIELDS OF GREAT BRITAIN.

A Text-Book of Agriculture. Adapted to the Syllabus of the Science and Art Department. For Elementary and Advanced Students. By HUGH CLEMENTS (Board of Trade). Second Edition, Revised, with Additions. 18mo, cloth. **2/6**

"It is a long time since we have seen a book which has pleased us more, or which contains such a vast and useful fund of knowledge."—*Educational Times.*

TABLES and MEMORANDA for FARMERS, GRAZIERS, AGRICULTURAL STUDENTS, SURVEYORS, LAND AGENTS, AUCTIONEERS, &c.

With a New System of Farm Book-keeping. By SIDNEY FRANCIS. Third Edition, Revised. 272 pp., waistcoat-pocket size, limp leather . . **1/6**

"Weighing less than 1 oz., and occupying no more space than a match-box, it contains a mass of facts and calculations which has never before, in such handy form, been obtainable. Every operation on the farm is dealt with. The work may be taken as thoroughly accurate, the whole of the tables having been revised by Dr. Fream. We cordially recommend it."—*Bell's Weekly Messenger.*

THE ROTHAMSTED EXPERIMENTS AND THEIR PRACTICAL LESSONS FOR FARMERS.

Part I. STOCK. Part II. CROPS. By C. J. R. TIPPER. Crown 8vo, cloth. [*Just Published.* **3/6**

"We have no doubt that the book will be welcomed by a large class of farmers and others interested in agriculture."—*Standard.*

AGRICULTURE, FARMING, GARDENING, &c.

FERTILISERS AND FEEDING STUFFS.
A Handbook for the Practical Farmer. By BERNARD DYER, D.Sc. (Lond.). With the Text of the Fertilisers and Feeding Stuffs Act of 1893, &c. Third Edition, Revised. Crown 8vo, cloth. [*Just Published.* **1/0**
"This little book is precisely what it professes to be—'A Handbook for the Practical Farmer.' Dr. Dyer has done farmers good service in placing at their disposal so much useful information in so intelligible a form."—*The Times*

BEES FOR PLEASURE AND PROFIT.
A Guide to the Manipulation of Bees, the Production of Honey, and the General Management of the Apiary. By G. GORDON SAMSON. With numerous Illustrations. Crown 8vo, cloth **1/0**

BOOK-KEEPING for FARMERS and ESTATE OWNERS.
A Practical Treatise, presenting, in Three Plans, a System adapted for all Classes of Farms. By JOHNSON M. WOODMAN, Chartered Accountant. Second Edition, Revised. Crown 8vo, cloth boards, **3/6**; or, cloth limp, **2/6**
"The volume is a capital study of a most important subject."—*Agricultural Gazette.*

WOODMAN'S YEARLY FARM ACCOUNT BOOK.
Giving Weekly Labour Account and Diary, and showing the Income and Expenditure under each Department of Crops, Live Stock, Dairy, &c., &c. With Valuation, Profit and Loss Account, and Balance Sheet at the End of the Year. By JOHNSON M. WOODMAN, Chartered Accountant. Second Edition. Folio, half-bound . . . . **7/6** *Net.*
"Contains every requisite form for keeping farm accounts readily and accurately."—*Agriculture.*

THE FORCING GARDEN.
Or, How to Grow Early Fruits, Flowers and Vegetables. With Plans and Estimates for Building Glasshouses, Pits and Frames. With Illustrations. By SAMUEL WOOD. Crown 8vo, cloth **3/6**
"A good book, containing a great deal of valuable teaching."—*Gardeners' Magazine.*

A PLAIN GUIDE TO GOOD GARDENING.
Or, How to Grow Vegetables, Fruits, and Flowers. By S. WOOD. Fourth Edition, with considerable Additions, and numerous Illustrations. Crown 8vo, cloth **3/6**
"A very good book, and one to be highly recommended as a practical guide. The practical directions are excellent."—*Athenæum.*

MULTUM-IN-PARVO GARDENING.
Or, How to Make One Acre of Land produce £620 a year, by the Cultivation of Fruits and Vegetables; also, How to Grow Flowers in Three Glass Houses, so as to realise £176 per annum clear Profit. By SAMUEL WOOD, Author of "Good Gardening," &c. Fifth and Cheaper Edition, Revised, with Additions. Crown 8vo, sewed **1/0**
"We are bound to recommend it as not only suited to the case of the amateur and gentleman's gardener, but to the market grower."—*Gardeners' Magazine.*

THE LADIES' MULTUM-IN-PARVO FLOWER GARDEN.
And Amateur's Complete Guide. By S. WOOD. Crown 8vo, cloth . **3/6**
"Full of shrewd hints and useful instructions, based on a lifetime of experience."—*Scotsman.*

POTATOES: HOW TO GROW AND SHOW THEM.
A Practical Guide to the Cultivation and General Treatment of the Potato. By J. PINK. Crown 8vo **2/0**

MARKET AND KITCHEN GARDENING.
By C. W. SHAW, late Editor of *Gardening Illustrated.* Cloth . . **3/6**
"The most valuable compendium of kitchen and market-garden work published."—*Farmer.*

46   CROSBY LOCKWOOD & SON'S CATALOGUE.

AUCTIONEERING, VALUING, LAND SURVEYING, ESTATE AGENCY, &c.

THE APPRAISER, AUCTIONEER, BROKER, HOUSE AND ESTATE AGENT AND VALUER'S POCKET ASSISTANT.
For the Valuation for Purchase, Sale, or Renewal of Leases, Annuities, and Reversions, and of Property generally; with Prices for Inventories, &c. By JOHN WHEELER, Valuer, &c. Sixth Edition, Re-written and greatly Extended by C. NORRIS, Surveyor, Valuer, &c. Royal 32mo, cloth . . . **5/0**
"A neat and concise book of reference, containing an admirable and clearly-arranged list of prices for inventories, and a very practical guide to determine the value of furniture, &c."—*Standard.*
"Contains a large quantity of varied and useful information as to the valuation for purchase, sale, or renewal of leases, annuities and reversions, and of property generally, with prices for inventories, and a guide to determine the value of interior fittings and other effects."—*Builder.*

AUCTIONEERS: THEIR DUTIES AND LIABILITIES.
A Manual of Instruction and Counsel for the Young Auctioneer. By ROBERT SQUIBBS, Auctioneer. Second Edition, Revised and partly Re-written. Demy 8vo, cloth **12/6**
"The standard text-book on the topics of which it treats."—*Athenæum.*
"The work is one of general excellent character, and gives much information in a compendious and satisfactory form."—*Builder.*
"May be recommended as giving a great deal of information on the law relating to auctioneers, in a very readable form."—*Law Journal.*
"Auctioneers may be congratulated on having so pleasing a writer to minister to their special needs."—*Solicitors' Journal.*

TABLES FOR THE PURCHASING OF ESTATES: FREEHOLD, COPYHOLD, OR LEASEHOLD; ANNUITIES, ADVOWSONS, &c.
And for the Renewing of Leases held under Cathedral Churches, Colleges, or other Corporate bodies, for Terms of Years certain, and for Lives; also for Valuing Reversionary Estates, Deferred Annuities, Next Presentations, &c.; together with SMART's Five Tables of Compound Interest, and an Extension of the same to Lower and Intermediate Rates. By W. INWOOD. 24th Edition, with considerable Additions, and new and valuable Tables of Logarithms for the more difficult Computations of the Interest of Money, Discount, Annuities, &c., by M. FÉDOR THOMAN. Crown 8vo, cloth **8/0**
"Those interested in the purchase and sale of estates, and in the adjustment of compensation cases, as well as in transactions in annuities, life insurances, &c., will find the present edition of eminent service."—*Engineering.*
"'Inwood's Tables' still maintain a most enviable reputation. The new issue has been enriched by large additional contributions by M. Fédor Thoman, whose carefully arranged Tables cannot fail to be of the utmost utility."—*Mining Journal.*

THE AGRICULTURAL VALUER'S ASSISTANT.
A Practical Handbook on the Valuation of Landed Estates; including Rules and Data for Measuring and Estimating the Contents, Weights and Values of Agricultural Produce and Timber, and the Values of Feeding Stuffs, Manures, and Labour; with Forms of Tenant-Right Valuations, Lists of Local Agricultural Customs, Scales of Compensation under the Agricultural Holdings Act, &c., &c. By TOM BRIGHT, Agricultural Surveyor. Second Edition, Enlarged. Crown 8vo, cloth **5/0**
"Full of tables and examples in connection with the valuation of tenant-right, estates, labour, contents and weights of timber, and farm produce of all kinds."—*Agricultural Gazette.*
"An eminently practical handbook, full of practical tables and data of undoubted interest and value to surveyors and auctioneers in preparing valuations of all kinds."—*Farmer.*

POLE PLANTATIONS AND UNDERWOODS.
A Practical Handbook on Estimating the Cost of Forming, Renovating, Improving, and Grubbing Plantations and Underwoods, their Valuation for Purposes of Transfer, Rental, Sale or Assessment. By TOM BRIGHT. Crown 8vo, cloth **3/6**
"To valuers, foresters and agents it will be a welcome aid."—*North British Agriculturist.*
"Well calculated to assist the valuer in the discharge of his duties, and of undoubted interest and use both to surveyors and auctioneers in preparing valuations of all kinds."—*Kent Herald.*

AUCTIONEERING, VALUING, LAND SURVEYING, &c. 47

THE LAND VALUER'S BEST ASSISTANT.
Being Tables on a very much Improved Plan, for Calculating the Value of Estates. With Tables for reducing Scotch, Irish, and Provincial Customary Acres to Statute Measure, &c. By R. HUDSON, C.E. New Edition. Royal 32mo, leather, elastic band **4/0**
"Of incalculable value to the country gentleman and professional man."—*Farmers' Journal.*

THE LAND IMPROVER'S POCKET-BOOK.
Comprising Formulæ, Tables, and Memoranda required in any Computation relating to the Permanent Improvement of Landed Property. By JOHN EWART, Surveyor. Second Edition, Revised. Royal 32mo, oblong, leather . **4/0**
"A compendious and handy little volume."—*Spectator.*

THE LAND VALUER'S COMPLETE POCKET-BOOK.
Being the above Two Works bound together. Leather **7/6**

HANDBOOK OF HOUSE PROPERTY.
A Popular and Practical Guide to the Purchase, Mortgage, Tenancy, and Compulsory Sale of Houses and Land, including the Law of Dilapidations and Fixtures: with Examples of all kinds of Valuations, Useful Information on Building and Suggestive Elucidations of Fine Art. By E. L. TARBUCK, Architect and Surveyor. Sixth Edition. 12mo, cloth . . . **5/0**

LAW AND MISCELLANEOUS.

MODERN JOURNALISM.
A Handbook of Instruction and Counsel for the Young Journalist. By JOHN B. MACKIE, Fellow of the Institute of Journalists. Crown 8vo, cloth . **2/0**
"This invaluable guide to journalism is a work which all aspirants to a journalistic career will read with advantage."—*Journalist.*

HANDBOOK FOR SOLICITORS AND ENGINEERS
Engaged in Promoting Private Acts of Parliament and Provisional Orders for the Authorisation of Railways, Tramways, Gas and Water Works, &c. By L. LIVINGSTONE MACASSEY, of the Middle Temple, Barrister-at-Law, M. Inst. C.E. 8vo, cloth . . . . **£1 5s.**

PATENTS for INVENTIONS, HOW to PROCURE THEM.
Compiled for the Use of Inventors, Patentees and others. By G. G. M. HARDINGHAM, Assoc. Mem. Inst. C.E., &c. Demy 8vo, cloth . . **1/6**

CONCILIATION & ARBITRATION in LABOUR DISPUTES.
A Historical Sketch and Brief Statement of the Present Position of the Question at Home and Abroad. By J. S. JEANS, Author of "England's Supremacy," &c. Crown 8vo, 200 pp., cloth **2/6**

THE HEALTH OFFICER'S POCKET-BOOK.
A Guide to Sanitary Practice and Law. For Medical Officers of Health, Sanitary Inspectors, Members of Sanitary Authorities, &c. By EDWARD F. WILLOUGHBY, M.D. (Lond.), &c. Fcap. 8vo, cloth . . . **7/6**
"A mine of condensed information of a pertinent and useful kind on the various subjects of which it treats. The matter seems to have been carefully compiled and arranged for facility of reference, and it is well illustrated by diagrams and woodcuts. The different subjects are succinctly but fully and scientifically dealt with."—*The Lancet.*
"Ought to be welcome to those for whose use it is designed, since it practically boils down a reference library into a pocket volume. . . . It combines, with an uncommon degree of efficiency, the qualities of accuracy, conciseness and comprehensiveness."—*Scotsman.*

EVERY MAN'S OWN LAWYER.

A Handy-Book of the Principles of Law and Equity. With a Concise Dictionary of Legal Terms. By A BARRISTER. Thirty-fifth Edition, carefully Revised, and including New Acts of Parliament of 1897. Comprising the *Workmen's Compensation Act, 1897; Voluntary Schools Acts, 1897; Preferential Payments in Bankruptcy Amendment Act, 1897; Weights and Measures (Metric System) Act, 1897; Infant Life Protection Act, 1897; Dangerous Performances (of Children) Act, 1897; Land Transfer Act, 1897, &c., &c. Judicial Decisions during the year have also been duly noted.* Crown 8vo, 750 pp. Price **6/8** (saved at every consultation!), strongly bound in cloth. *[Just Published.*

The Book will be found to comprise (amongst other matter)—

THE RIGHTS AND WRONGS OF INDIVIDUALS—LANDLORD AND TENANT—VENDORS AND PURCHASERS—LEASES AND MORTGAGES—PRINCIPAL AND AGENT—PARTNERSHIP AND COMPANIES—MASTERS, SERVANTS AND WORKMEN—CONTRACTS AND AGREEMENTS—BORROWERS, LENDERS AND SURETIES—SALE AND PURCHASE OF GOODS—CHEQUES, BILLS AND NOTES—BILLS OF SALE—BANKRUPTCY—RAILWAY AND SHIPPING LAW—LIFE, FIRE, AND MARINE INSURANCE—ACCIDENT AND FIDELITY INSURANCE—CRIMINAL LAW—PARLIAMENTARY ELECTIONS—COUNTY COUNCILS—DISTRICT COUNCILS—PARISH COUNCILS—MUNICIPAL CORPORATIONS—LIBEL AND SLANDER—PUBLIC HEALTH AND NUISANCES—COPYRIGHT, PATENTS, TRADE MARKS—HUSBAND AND WIFE—DIVORCE—INFANCY—CUSTODY OF CHILDREN—TRUSTEES AND EXECUTORS—CLERGY, CHURCH-WARDENS, &c.—GAME LAWS AND SPORTING—INNKEEPERS—HORSES AND DOGS—TAXES AND DEATH DUTIES—FORMS OF AGREEMENTS, WILLS, CODICILS, NOTICES, &c.

☞ *The object of this work is to enable those who consult it to help themselves to the law; and thereby to dispense, as far as possible, with professional assistance and advice. There are many wrongs and grievances which persons submit to from time to time through not knowing how or where to apply for redress; and many persons have as great a dread of a lawyer's office as of a lion's den. With this book at hand it is believed that many a SIX-AND-EIGHTPENCE may be saved; many a wrong redressed; many a right reclaimed; many a law suit avoided; and many an evil abated. The work has established itself as the standard legal adviser of all classes, and has also made a reputation for itself as a useful book of reference for lawyers residing at a distance from law libraries, who are glad to have at hand a work embodying recent decisions and enactments.*

OPINIONS OF THE PRESS.

"It is a complete code of English Law written in plain language, which all can understand. Should be in the hands of every business man, and all who wish to abolish lawyers' bills."—*Weekly Times.*

"A useful and concise epitome of the law, compiled with considerable care."—*Law Magazine.*

"A complete digest of the most useful facts which constitute English law."—*Globe.*

"This excellent handbook. . . . Admirably done, admirably arranged, and admirably cheap."—*Leeds Mercury.*

"A concise, cheap, and complete epitome of the English law. So plainly written that he who runs may read, and he who reads may understand."—*Figaro.*

"A dictionary of legal facts well put together. The book is a very useful one."—*Spectator.*

THE PAWNBROKER'S, FACTOR'S, AND MERCHANT'S GUIDE TO THE LAW OF LOANS AND PLEDGES.

With the Statutes and a Digest of Cases. By H. C. FOLKARD, Barrister-at-Law. Cloth **3/6**

LABOUR CONTRACTS.

A Popular Handbook on the Law of Contracts for Works and Services. By DAVID GIBBONS. Fourth Edition, with Appendix of Statutes by T. F. UTTLEY, Solicitor. Fcap. 8vo, cloth **3/6**

SUMMARY OF THE FACTORY AND WORKSHOP ACTS

(1878-1891). For the Use of Manufacturers and Managers. By EMILE GARCKE and J. M. FELLS. (Reprinted from "FACTORY ACCOUNTS.") Crown 8vo, sewed . . . . **6D.**

WEALE'S SERIES

OF

SCIENTIFIC AND TECHNICAL WORKS.

"It is not too much to say that no books have ever proved more popular with or more useful to young engineers and others than the excellent treatises comprised in WEALE'S SERIES."—*Engineer*.

A New Classified List.

	PAGE		PAGE
CIVIL ENGINEERING AND SURVEYING	2	ARCHITECTURE AND BUILDING . .	6
MINING AND METALLURGY . . .	3	INDUSTRIAL AND USEFUL ARTS. .	8
MECHANICAL ENGINEERING . . .	4	AGRICULTURE, GARDENING, ETC. .	10
NAVIGATION, SHIPBUILDING, ETC. .	5	MATHEMATICS, ARITHMETIC, ETC. .	12
BOOKS OF REFERENCE AND MISCELLANEOUS VOLUMES		14	

CROSBY LOCKWOOD AND SON,
7, STATIONERS' HALL COURT, LONDON, E.C.
1899.

WEALE'S SCIENTIFIC AND TECHNICAL SERIES.

CIVIL ENGINEERING & SURVEYING.

Civil Engineering.
By HENRY LAW, M. Inst. C.E. Including a Treatise on HYDRAULIC ENGINEERING by G. R. BURNELL, M.I.C.E. Seventh Edition, revised, with LARGE ADDITIONS by D. K. CLARK, M.I.C.E. . . . **6/6**

Pioneer Engineering:
A Treatise on the Engineering Operations connected with the Settlement of Waste Lands in New Countries. By EDWARD DOBSON, A.I.C.E. With numerous Plates. Second Edition **4/6**

Iron Bridges of Moderate Span:
Their Construction and Erection. By HAMILTON W. PENDRED. With 40 Illustrations **2/0**

Iron and Steel Bridges and Viaducts.
A Practical Treatise upon their Construction for the use of Engineers, Draughtsmen, and Students. By FRANCIS CAMPIN, C.E. With numerous Illustrations . . . . [*Just Published* **3/6**

Constructional Iron and Steel Work,
As applied to Public, Private, and Domestic Buildings. By FRANCIS CAMPIN, C.E. **3/6**

Tubular and other Iron Girder Bridges.
Describing the Britannia and Conway Tubular Bridges. By G. DRYSDALE DEMPSEY, C.E. Fourth Edition **2/0**

Materials and Construction:
A Theoretical and Practical Treatise on the Strains, Designing, and Erection of Works of Construction. By FRANCIS CAMPIN, C.E. . . **3/0**

Sanitary Work in the Smaller Towns and in Villages.
By CHARLES SLAGG, Assoc. M. Inst. C.E. Second Edition . . **3/0**

Roads and Streets (The Construction of).
In Two Parts: I. THE ART OF CONSTRUCTING COMMON ROADS, by H. LAW, C.E., Revised by D. K. CLARK, C.E.; II. RECENT PRACTICE: Including Pavements of Wood, Asphalte, etc. By D. K. CLARK, C.E. **4/6**

Gas Works (The Construction of),
And the Manufacture and Distribution of Coal Gas. By S. HUGHES, C.E. Re-written by WILLIAM RICHARDS, C.E. Eighth Edition . . **5/6**

Water Works
For the Supply of Cities and Towns. With a Description of the Principal Geological Formations of England as influencing Supplies of Water. By SAMUEL HUGHES, F.G.S., C.E. Enlarged Edition **4/0**

The Power of Water,
As applied to drive Flour Mills, and to give motion to Turbines and other Hydrostatic Engines. By JOSEPH GLYNN, F.R.S. New Edition . **2/0**

Wells and Well-Sinking.
By JOHN GEO. SWINDELL, A.R.I.B.A., and G. R. BURNELL, C.E. Revised Edition. With a New Appendix on the Qualities of Water. Illustrated **2/0**

The Drainage of Lands, Towns, and Buildings.
By G. D. DEMPSEY, C.E. Revised, with large Additions on Recent Practice, by D K. CLARK, M.I.C.E. Third Edition . . . **4/6**

Embanking Lands from the Sea.
With Particulars of actual Embankments, &c. By JOHN WIGGINS . **2/0**

The Blasting and Quarrying of Stone,
For Building and other Purposes. With Remarks on the Blowing up of Bridges. By Gen. Sir J. BURGOYNE, K.C B. **1/6**

Foundations and Concrete Works.
With Practical Remarks on Footings, Planking, Sand, Concrete, Béton, Pile-driving, Caissons, and Cofferdams. By E. DOBSON, M.R.I.B.A. Seventh Edition **1/6**

WEALE'S SCIENTIFIC AND TECHNICAL SERIES. 3

Pneumatics,
Including Acoustics and the Phenomena of Wind Currents, for the Use of Beginners. By CHARLES TOMLINSON, F.R.S. Fourth Edition . **1/6**

Land and Engineering Surveying.
For Students and Practical Use. By T. BAKER, C.E. Seventeenth Edition, Revised and Extended by F. E. DIXON, A.M. Inst. C.E., Professional Associate of the Institution of Surveyors. With numerous Illustrations and two Lithographic Plates . . . . . . *[Just published* **2/0**

Mensuration and Measuring.
For Students and Practical Use. With the Mensuration and Levelling of Land for the purposes of Modern Engineering. By T. BAKER, C.E. New Edition by E. NUGENT, C.E. **1/6**

MINING AND METALLURGY

Mineralogy,
Rudiments of. By A. RAMSAY, F.G.S. Third Edition, revised and enlarged. Woodcuts and Plates **3/6**

Coal and Coal Mining,
A Rudimentary Treatise on By the late Sir WARINGTON W. SMYTH, F.R.S. Seventh Edition, revised and enlarged **3/6**

Metallurgy of Iron.
Containing Methods of Assay, Analyses of Iron Ores, Processes of Manufacture of Iron and Steel, &c. By H. BAUERMAN, F.G.S. With numerous Illustrations. Sixth Edition, revised and enlarged . . . **5/0**

The Mineral Surveyor and Valuer's Complete Guide.
By W. LINTERN. Fourth Edition, with an Appendix on Magnetic and Angular Surveying **3/6**

Slate and Slate Quarrying:
Scientific, Practical, and Commercial. By D. C. DAVIES, F.G.S. With numerous Illustrations and Folding Plates. Third Edition . . **3/0**

A First Book of Mining and Quarrying,
With the Sciences connected therewith, for Primary Schools and Self Instruction. By J. H. COLLINS, F.G.S. Second Edition . . . **1/6**

Subterraneous Surveying,
With and without the Magnetic Needle. By T. FENWICK and T. BAKER, C.E. Illustrated **2/6**

Mining Tools.
Manual of. By WILLIAM MORGANS, Lecturer on Practical Mining at the Bristol School of Mines **2/6**

Mining Tools, Atlas
Of Engravings to Illustrate the above, containing 235 Illustrations of Mining Tools, drawn to Scale. 4to **4/6**

Physical Geology,
Partly based on Major-General PORTLOCK'S "Rudiments of Geology." By RALPH TATE, A.L.S., &c. Woodcuts. **2/0**

Historical Geology,
Partly based on Major-General PORTLOCK'S "Rudiments." By RALPH TATE, A.L.S., &c. Woodcuts **2/6**

Geology, Physical and Historical.
Consisting of "Physical Geology," which sets forth the Leading Principles of the Science; and "Historical Geology," which treats of the Mineral and Organic Conditions of the Earth at each successive epoch. By RALPH TATE, F.G.S. **4/6**

Electro-Metallurgy,
Practically Treated. By ALEXANDER WATT. Tenth Edition, enlarged and revised, including the most Recent Processes . . . **3/6**

MECHANICAL ENGINEERING.

The Workman's Manual of Engineering Drawing.
By JOHN MAXTON, Instructor in Engineering Drawing, Royal Naval College, Greenwich. Seventh Edition. 300 Plates and Diagrams . **3/6**

Fuels: Solid, Liquid, and Gaseous.
Their Analysis and Valuation. For the Use of Chemists and Engineers. By H. J. PHILLIPS, F.C.S., formerly Analytical and Consulting Chemist to the Great Eastern Railway. Second Edition, Revised. . . **2/0**

Fuel, Its Combustion and Economy.
Consisting of an Abridgment of "A Treatise on the Combustion of Coal and the Prevention of Smoke." By C. W. WILLIAMS, A.I.C.E. With Extensive Additions by D. K. CLARK, M. Inst. C.E. Third Edition . **3/6**

The Boilermaker's Assistant
In Drawing, Templating, and Calculating Boiler Work, &c. By J. COURTNEY, Practical Boilermaker. Edited by D. K. CLARK, C.E. . **2/0**

The Boiler-Maker's Ready Reckoner,
With Examples of Practical Geometry and Templating for the Use of Platers, Smiths, and Riveters. By JOHN COURTNEY. Edited by D. K. CLARK, M.I.C.E. Second Edition, revised, with Additions . . **4/0**

*** *The last two Works in One Volume, half-bound, entitled* "THE BOILERMAKER'S READY-RECKONER AND ASSISTANT." By J. COURTNEY and D. K. CLARK. *Price 7s.*

Steam Boilers:
Their Construction and Management. By R. ARMSTRONG, C.E. Illustrated **1/6**

Steam and Machinery Management.
A Guide to the Arrangement and Economical Management of Machinery. By M. POWIS BALE, M. Inst. M.E. **2/6**

Steam and the Steam Engine,
Stationary and Portable. Being an Extension of the Treatise on the Steam Engine of Mr. J. SEWELL. By D. K. CLARK, C.E. Third Edition **3/6**

The Steam Engine,
A Treatise on the Mathematical Theory of, with Rules and Examples for Practical Men. By T. BAKER, C.E. **1/6**

The Steam Engine.
By Dr. LARDNER. Illustrated **1/6**

Locomotive Engines,
By G. D. DEMPSEY, C.E. With large Additions treating of the Modern Locomotive, by D. K. CLARK, M. Inst. C.E. **3/0**

Locomotive Engine-Driving.
A Practical Manual for Engineers in charge of Locomotive Engines. By MICHAEL REYNOLDS. Eighth Edition. 3s. 6d. limp; cloth boards **4/6**

Stationary Engine-Driving.
A Practical Manual for Engineers in charge of Stationary Engines. By MICHAEL REYNOLDS. Fourth Edition. 3s. 6d. limp; cloth boards . **4/6**

The Smithy and Forge.
Including the Farrier's Art and Coach Smithing. By W. J. E. CRANE. Second Edition, revised **2/6**

Modern Workshop Practice,
As applied to Marine, Land, and Locomotive Engines, Floating Docks, Dredging Machines, Bridges, Ship-building, &c. By J. G. WINTON. Fourth Edition, Illustrated **3/6**

Mechanical Engineering.
Comprising Metallurgy, Moulding, Casting, Forging, Tools, Workshop Machinery, Mechanical Manipulation, Manufacture of the Steam Engine, &c. By FRANCIS CAMPIN, C.E. Third Edition **2/6**

Details of Machinery.
Comprising Instructions for the Execution of various Works in Iron in the Fitting-Shop, Foundry, and Boiler-Yard. By FRANCIS CAMPIN, C.E. **3/0**

Elementary Engineering:
A Manual for Young Marine Engineers and Apprentices. In the Form of Questions and Answers on Metals, Alloys, Strength of Materials, &c. By J. S. BREWER. Second Edition **2/0**

Power in Motion:
Horse-power Motion, Toothed-Wheel Gearing, Long and Short Driving Bands, Angular Forces, &c. By JAMES ARMOUR, C.E. Third Edition **2/0**

Iron and Heat,
Exhibiting the Principles concerned in the Construction of Iron Beams, Pillars, and Girders. By J. ARMOUR, C.E. **2/6**

Practical Mechanism,
And Machine Tools. By T. BAKER, C.E. With Remarks on Tools and Machinery, by J. NASMYTH, C.E. **2/6**

Mechanics:
Being a concise Exposition of the General Principles of Mechanical Science, and their Applications. By CHARLES TOMLINSON, F.R.S. . . **1/6**

Cranes (The Construction of),
And other Machinery for Raising Heavy Bodies for the Erection of Buildings, &c. By JOSEPH GLYNN, F.R.S. **1/6**

NAVIGATION, SHIPBUILDING, ETC.

The Sailor's Sea Book:
A Rudimentary Treatise on Navigation. By JAMES GREENWOOD, B.A. With numerous Woodcuts and Coloured Plates. New and enlarged Edition. By W. H. ROSSER **2/6**

Practical Navigation.
Consisting of THE SAILOR'S SEA-BOOK, by JAMES GREENWOOD and W. H. ROSSER; together with Mathematical and Nautical Tables for the Working of the Problems, by HENRY LAW, C.E., and Prof. J. R. YOUNG . **7/0**

Navigation and Nautical Astronomy,
In Theory and Practice. By Prof. J. R. YOUNG. New Edition. **2/6**

Mathematical Tables,
For Trigonometrical, Astronomical, and Nautical Calculations; to which is prefixed a Treatise on Logarithms. By H. LAW, C.E. Together with a Series of Tables for Navigation and Nautical Astronomy. By Professor J. R. YOUNG. New Edition **4/0**

Masting, Mast-Making, and Rigging of Ships.
Also Tables of Spars, Rigging, Blocks; Chain, Wire, and Hemp Ropes, &c., relative to every class of vessels. By ROBERT KIPPING, N.A. . **2 0**

Sails and Sail-Making.
With Draughting, and the Centre of Effort of the Sails. By ROBERT KIPPING, N.A. **2/6**

Marine Engines and Steam Vessels.
By R. MURRAY, C.E. Eighth Edition, thoroughly revised, with Additions by the Author and by GEORGE CARLISLE. C E. . . . **4/6**

Iron Ship-Building.
With Practical Examples. By JOHN GRANTHAM. Fifth Edition . **4/0**

Naval Architecture:
An Exposition of Elementary Principles. By JAMES PEAKE . . **3 6**

Ships for Ocean and River Service,
Principles of the Construction of. By HAKON A. SOMMERFELDT . **1 '6**

Atlas of Engravings
To Illustrate the above. Twelve large folding Plates. Royal 4to, cloth **7 6**

The Forms of Ships and Boats.
By W. BLAND. Seventh Edition, revised, with numerous Illustrations and Models. **1/6**

ARCHITECTURE AND THE BUILDING ARTS.

Constructional Iron and Steel Work,
As applied to Public, Private, and Domestic Buildings. By FRANCIS CAMPIN, C.E. **3/6**

Building Estates:
A Treatise on the Development, Sale, Purchase, and Management of Building Land. By F. MAITLAND. Second Edition, revised . . **2/0**

The Science of Building:
An Elementary Treatise on the Principles of Construction. By E. WYNDHAM TARN, M.A. Lond. Third Edition, revised and enlarged . **3/6**

The Art of Building:
General Principles of Construction, Strength, and Use of Materials, Working Drawings, Specifications, &c. By EDWARD DOBSON, M.R.I.B.A. . **2/0**

A Book on Building,
Civil and Ecclesiastical. By Sir EDMUND BECKETT, Q.C. (Lord GRIMTHORPE). Second Edition **4/6**

Dwelling-Houses (The Erection of),
Illustrated by a Perspective View, Plans, and Sections of a Pair of Villas, with Specification, Quantities, and Estimates. By S. H. BROOKS, Architect **2/6**

Cottage Building.
By C. BRUCE ALLEN. Eleventh Edition, with Chapter on Economic Cottages for Allotments, by E. E. ALLEN, C.E. **2/0**

Acoustics in Relation to Architecture and Building:
The Laws of Sound as applied to the Arrangement of Buildings. By Professor T. ROGER SMITH, F.R.I.B.A. New Edition, Revised . **1/6**

The Rudiments of Practical Bricklaying.
General Principles of Bricklaying; Arch Drawing, Cutting, and Setting; Pointing; Paving, Tiling, &c. By ADAM HAMMOND. With 68 Woodcuts **1/6**

The Art of Practical Brick Cutting and Setting.
By ADAM HAMMOND. With 9 Engravings **1/6**

Brickwork:
A Practical Treatise, embodying the General and Higher Principles of Bricklaying, Cutting and Setting; with the Application of Geometry to Roof Tiling, &c. By F. WALKER **1/6**

Bricks and Tiles,
Rudimentary Treatise on the Manufacture of; containing an Outline of the Principles of Brickmaking. By E. DOBSON, M.R.I.B.A. Additions by C. TOMLINSON, F.R.S. Illustrated **3/0**

The Practical Brick and Tile Book.
Comprising: BRICK AND TILE MAKING, by E. DOBSON, A.I.C.E.; Practical BRICKLAYING, by A. HAMMOND; BRICK-CUTTING AND SETTING, by A. HAMMOND. 550 pp. with 270 Illustrations, half-bound . . **6/0**

Carpentry and Joinery—
THE ELEMENTARY PRINCIPLES OF CARPENTRY. Chiefly composed from the Standard Work of THOMAS TREDGOLD, C.E. With Additions, and TREATISE ON JOINERY, by E. W. TARN, M.A. Sixth Edition, Revised . . **3/6**

Carpentry and Joinery—Atlas
Of 35 Plates to accompany and Illustrate the foregoing book. With Descriptive Letterpress. 4to **6/0**

A Practical Treatise on Handrailing;
Showing New and Simple Methods. By GEO. COLLINGS. Second Edition,
Revised, including a TREATISE ON STAIRBUILDING. With Plates . **2/6**

Circular Work in Carpentry and Joinery.
A Practical Treatise on Circular Work of Single and Double Curvature.
By GEORGE COLLINGS. Second Edition **2/6**

Roof Carpentry:
Practical Lessons in the Framing of Wood Roofs. For the Use of Working
Carpenters. By GEO. COLLINGS **2/0**

The Construction of Roofs of Wood and Iron;
Deduced chiefly from the Works of Robison, Tredgold, and Humber. By
E. WYNDHAM TARN, M.A., Architect. Second Edition, revised . **1/6**

The Joints Made and Used by Builders.
By WYVILL J. CHRISTY, Architect. With 160 Woodcuts . . **3/0**

Shoring
And its Application: A Handbook for the Use of Students. By GEORGE
H. BLAGROVE. With 31 Illustrations **1/6**

The Timber Importer's, Timber Merchant's, and Builder's Standard Guide.
By R. E. GRANDY **2/0**

Plumbing:
A Text-Book to the Practice of the Art or Craft of the Plumber. With
Chapters upon House Drainage and Ventilation. By WM. PATON BUCHAN.
Seventh Edition, Re-written and Enlarged, with 500 Illustrations . **3/6**

Ventilation:
A Text Book to the Practice of the Art of Ventilating Buildings. By W. P.
BUCHAN, R.P., Author of "Plumbing," &c. With 170 Illustrations **3/6**

The Practical Plasterer:
A Compendium of Plain and Ornamental Plaster Work. By W. KEMP **2/0**

House Painting, Graining, Marbling, & Sign Writing.
With a Course of Elementary Drawing, and a Collection of Useful Receipts.
By ELLIS A. DAVIDSON. Seventh Edition. Coloured Plates . . **5/0**
*** *The above, in cloth boards, strongly bound, 6s.*

A Grammar of Colouring,
Applied to Decorative Painting and the Arts. By GEORGE FIELD. New
Edition, enlarged, by ELLIS A. DAVIDSON. With Coloured Plates . **3/0**

Elementary Decoration
As applied to Dwelling Houses, &c. By JAMES W. FACEY. Illustrated **2/0**

Practical House Decoration.
A Guide to the Art of Ornamental Painting, the Arrangement of Colours in
Apartments, and the Principles of Decorative Design. By JAMES W. FACEY. **2/6**
*** *The last two Works in One handsome Vol., half-bound, entitled* "HOUSE
DECORATION, ELEMENTARY AND PRACTICAL," *price 5s.*

Warming and Ventilation
Of Domestic and Public Buildings, Mines, Lighthouses, Ships, &c. By
CHARLES TOMLINSON, F.R.S. **3/0**

Portland Cement for Users.
By HENRY FAIJA, A.M. Inst. C.E. Third Edition, Corrected . **2/0**

Limes, Cements, Mortars, Concretes, Mastics, Plastering, &c.
By G. R. BURNELL, C.E. Thirteenth Edition . . . **1/6**

Masonry and Stone-Cutting.
The Principles of Masonic Projection and their application to Construction. By EDWARD DOBSON, M.R.I.B.A. **2/6**

Arches, Piers, Buttresses, &c.:
Experimental Essays on the Principles of Construction. By W. BLAND. **1/6**

Quantities and Measurements,
In Bricklayers', Masons', Plasterers', Plumbers', Painters', Paperhangers', Gilders', Smiths', Carpenters' and Joiners' Work. By A. C. BEATON **1/6**

The Complete Measurer:
Setting forth the Measurement of Boards, Glass, Timber and Stone. By R. HORTON. Fifth Edition **4/0**

** *The above, strongly bound in leather, price 5s.*

Light:
An Introduction to the Science of Optics. Designed for the Use of Students of Architecture, Engineering, and other Applied Sciences. By E. WYNDHAM TARN, M.A., Author of "The Science of Building," &c. . . **1/6**

Hints to Young Architects.
By GEORGE WIGHTWICK, Architect. Fifth Edition, revised and enlarged by G. HUSKISSON GUILLAUME, Architect **3/6**

Architecture—Orders:
The Orders and their Æsthetic Principles. By W. H. LEEDS. Illustrated. **1/6**

Architecture—Styles:
The History and Description of the Styles of Architecture of Various Countries, from the Earliest to the Present Period. By T. TALBOT BURY, F.R.I.B.A. Illustrated **2/0**

** ORDERS AND STYLES OF ARCHITECTURE, *in One Vol., 3s. 6d.*

Architecture—Design:
The Principles of Design in Architecture, as deducible from Nature and exemplified in the Works of the Greek and Gothic Architects. By EDW. LACY GARBETT, Architect. Illustrated **2/6**

** *The three preceding Works in One handsome Vol., half bound, entitled* "MODERN ARCHITECTURE," *price 6s.*

Perspective for Beginners.
Adapted to Young Students and Amateurs in Architecture, Painting, &c. By GEORGE PYNE **2/0**

Architectural Modelling in Paper.
By T. A. RICHARDSON. With Illustrations, engraved by O. JEWITT **1/6**

Glass Staining, and the Art of Painting on Glass.
From the German of Dr. GESSERT and EMANUEL OTTO FROMBERG. With an Appendix on THE ART OF ENAMELLING **2/6**

Vitruvius—The Architecture of.
In Ten Books. Translated from the Latin by JOSEPH GWILT, F.S.A., F.R.A.S. With 23 Plates **5/0**

N.B.—This is the only Edition of VITRUVIUS *procurable at a moderate price.*

Grecian Architecture,
An Inquiry into the Principles of Beauty in. With an Historical View of the Rise and Progress of the Art in Greece. By the EARL OF ABERDEEN **1/0**

** *The two preceding Works in One handsome Vol., half bound, entitled* "ANCIENT ARCHITECTURE." *price 6s.*

INDUSTRIAL AND USEFUL ARTS.

Cements, Pastes, Glues, and Gums.
A Practical Guide to the Manufacture and Application of the various Agglutinants required for Workshop, Laboratory, or Office Use. With upwards of 900 Recipes and Formulæ. By H. C. STANDAGE . . **2/0**

Clocks and Watches, and Bells,
A Rudimentary Treatise on. By Sir EDMUND BECKETT, Q.C. (Lord GRIMTHORPE). Seventh Edition **4/6**

The Goldsmith's Handbook.
Containing full Instructions in the Art of Alloying, Melting, Reducing, Colouring, Collecting and Refining, Recovery of Waste, Solders, Enamels. &c., &c. By GEORGE E. GEE. Third Edition, enlarged . . . **3/0**

The Silversmith's Handbook,
On the same plan as the GOLDSMITH'S HANDBOOK. By GEORGE E. GEE. Second Edition, Revised **3/0**
*** *The lost two Works, in One handsome Vol., half-bound, 7s.*

The Hall-Marking of Jewellery.
Comprising an account of all the different Assay Towns of the United Kingdom; with the Stamps and Laws relating to the Standards and Hall-Marks at the various Assay Offices. By GEORGE E. GEE . . **3/0**

Practical Organ Building.
By W. E. DICKSON, M.A. Second Edition, Revised, with Additions **2/6**

Coach-Building:
A Practical Treatise. By JAMES W. BURGESS. With 57 Illustrations **2/6**

The Brass Founder's Manual:
Instructions for Modelling, Pattern Making, Moulding, Turning, &c. By W. GRAHAM **2/0**

The Sheet-Metal Worker's Guide.
A Practical Handbook for Tinsmiths, Coppersmiths, Zincworkers, &c., with 46 Diagrams. By W. J. E. CRANE. Second Edition, revised . **1/6**

Sewing Machinery:
Its Construction, History, &c. With full Technical Directions for Adjusting, &c. By J. W. URQUHART, C.E. **2/0**

Gas Fitting:
A Practical Handbook. By JOHN BLACK. Second Edition, Enlarged. With 130 Illustrations **2/6**

Construction of Door Locks.
From the Papers of A. C. HOBBS. Edited by CHARLES TOMLINSON, F.R.S. With a Note upon IRON SAFES by ROBERT MALLET. Illustrated . **2/6**

The Model Locomotive Engineer, Fireman, and Engine-Boy.
Comprising an Historical Notice of the Pioneer Locomotive Engines and their Inventors. By MICHAEL REYNOLDS. Second Edition. With numerous Illustrations, and Portrait of George Stephenson . . **3/6**

The Art of Letter Painting made Easy.
By J. G. BADENOCH. With 12 full-page Engravings of Examples . **1/6**

The Art of Boot and Shoemaking.
Including Measurement, Last-fitting, Cutting-out, Closing and Making. By JOHN BEDFORD LENO. With numerous Illustrations. Third Edition **2/0**

Mechanical Dentistry:
A Practical Treatise on the Construction of the Various Kinds of Artificial Dentures. By CHARLES HUNTER. Third Edition, revised . . **3/0**

Wood Engraving:
A Practical and Easy Introduction to the Art. By W. N. BROWN . **1/6**

Laundry Management.
A Handbook for Use in Private and Public Laundries. Including Accounts of Modern Machinery and Appliances. By the EDITOR of "The Laundry Journal." With numerous Illustrations. Second Edition . . **2/0**

AGRICULTURE, GARDENING, ETC.

Draining and Embanking:
A Practical Treatise. By Prof. JOHN SCOTT. With 68 Illustrations **1/6**

Irrigation and Water Supply:
A Practical Treatise on Water Meadows, Sewage Irrigation, Warping, &c.; on the Construction of Wells, Ponds, Reservoirs, &c. By Prof. JOHN SCOTT. With 34 Illustrations **1/6**

Farm Roads, Fences, and Gates:
A Practical Treatise on the Roads, Tramways, and Waterways of the Farm; the Principles of Enclosures; and the different kinds of Fences, Gates, and Stiles. By Prof. JOHN SCOTT. With 75 Illustrations . **1/6**

Farm Buildings:
A Practical Treatise on the Buildings necessary for various kinds of Farms, their Arrangement and Construction, with Plans and Estimates. By Prof. JOHN SCOTT. With 105 Illustrations **2/0**

Barn Implements and Machines:
Treating of the Application of Power and Machines used in the Threshing-barn, Stockyard, Dairy, &c. By Prof. J. SCOTT. With 123 Illustrations. **2/0**

Field Implements and Machines:
With Principles and Details of Construction and Points of Excellence, their Management, &c. By Prof. JOHN SCOTT. With 138 Illustrations **2/0**

Agricultural Surveying:
A Treatise on Land Surveying, Levelling, and Setting-out; with Directions for Valuing Estates. By Prof. J. SCOTT. With 62 Illustrations . **1/6**

Farm Engineering.
By Professor JOHN SCOTT. Comprising the above Seven Volumes in One, 1,150 pages, and over 600 Illustrations. Half-bound . . . **12/0**

Outlines of Farm Management.
Treating of the General Work of the Farm; Stock; Contract Work; Labour, &c. By R. SCOTT BURN **2/6**

Outlines of Landed Estates Management.
Treating of the Varieties of Lands, Methods of Farming, Setting-out of Farms, Roads, Fences, Gates, Drainage, &c. By R. SCOTT BURN. **2/6**

*** *The above Two Vols. in One, handsomely half-bound, price* **6s.**

Soils, Manures, and Crops.
(Vol. I. OUTLINES OF MODERN FARMING.) By R. SCOTT BURN . **2/0**

Farming and Farming Economy.
(Vol. II. OUTLINES OF MODERN FARMING.) By R. SCOTT BURN **3/0**

Stock: Cattle, Sheep, and Horses.
(Vol. III. OUTLINES OF MODERN FARMING.) By R. SCOTT BURN **2/6**

Dairy, Pigs, and Poultry.
(Vol. IV. OUTLINES OF MODERN FARMING.) By R. SCOTT BURN **2/0**

Utilization of Sewage, Irrigation, and Reclamation of Waste Land.
(Vol. V. OUTLINES OF MODERN FARMING.) By R. SCOTT BURN . **2/6**

Outlines of Modern Farming.
By R. SCOTT BURN. Consisting of the above Five Volumes in One, 1,250 pp., profusely Illustrated, half-bound **12/0**

WEALE'S SCIENTIFIC AND TECHNICAL SERIES. 11

Book-keeping for Farmers and Estate Owners.
A Practical Treatise, presenting, in Three Plans, a System adapted for all classes of Farms. By J. M. WOODMAN. Third Edition, revised . **2/6**

Ready Reckoner for the Admeasurement of Land.
By A. ARMAN. Third Edition, revised and extended by C. NORRIS **2/0**

Miller's, Corn Merchant's, and Farmer's Ready Reckoner.
Second Edition, revised, with a Price List of Modern Flour Mill Machinery, by W. S. HUTTON, C.E. **2/0**

The Hay and Straw Measurer.
New Tables for the Use of Auctioneers, Valuers, Farmers, Hay and Straw Dealers, &c. By JOHN STEELE **2/0**

Meat Production.
A Manual for Producers, Distributors, and Consumers of Butchers' Meat. By JOHN EWART **2/6**

Sheep:
The History, Structure, Economy, and Diseases of. By W. C. SPOONER, M.R.V.S. Fifth Edition, with fine Engravings. **3/6**

Market and Kitchen Gardening.
By C. W. SHAW, late Editor of "Gardening Illustrated" . . **3/0**

Kitchen Gardening Made Easy.
Showing the best means of Cultivating every known Vegetable and Herb, &c., with directions for management all the year round. By GEORGE M. F. GLENNY. Illustrated **1/6**

Cottage Gardening:
Or Flowers, Fruits, and Vegetables for Small Gardens. By E. HOBDAY.
1/6

Garden Receipts.
Edited by CHARLES W. QUIN . **1/6**

Fruit Trees,
The Scientific and Profitable Culture of. From the French of M. DU BREUIL. Fourth Edition, carefully Revised by GEORGE GLENNY. With 187 Woodcuts **3/6**

The Tree Planter and Plant Propagator:
With numerous Illustrations of Grafting, Layering, Budding, Implements, Houses, Pits, &c. By SAMUEL WOOD **2/0**

The Tree Pruner:
A Practical Manual on the Pruning of Fruit Trees, Shrubs, Climbers, and Flowering Plants. With numerous Illustrations. By SAMUEL WOOD **1/6**

*** *The above Two Vols. in One, handsomely half-bound, price* 3s. 6d.

The Art of Grafting and Budding.
By CHARLES BALTET. With Illustrations **2/6**

12 WEALE'S SCIENTIFIC AND TECHNICAL SERIES.

MATHEMATICS, ARITHMETIC, ETC.

Descriptive Geometry,
An Elementary Treatise on; with a Theory of Shadows and of Perspective, extracted from the French of G. MONGE. To which is added a Description of the Principles and Practice of Isometrical Projection. By J. F. HEATHER, M.A. With 14 Plates **2/0**

Practical Plane Geometry:
Giving the Simplest Modes of Constructing Figures contained in one Plane and Geometrical Construction of the Ground. By J. F. HEATHER, M.A. With 215 Woodcuts **2/0**

Analytical Geometry and Conic Sections,
A Rudimentary Treatise on. By JAMES HANN. A New Edition, re-written and enlarged by Professor J. R. YOUNG **2/0**

Euclid (The Elements of).
With many Additional Propositions and Explanatory Notes; to which is prefixed an Introductory Essay on Logic. By HENRY LAW, C.E. . **2/6**

* Sold also separately, viz:—
Euclid. The First Three Books. By HENRY LAW, C.E. . . **1/6**
Euclid. Books 4, 5, 6, 11, 12. By HENRY LAW, C.E. . . **1/6**

Plane Trigonometry,
The Elements of. By JAMES HANN. . . **1/6**

Spherical Trigonometry,
The Elements of. By JAMES HANN. Revised by CHARLES H. DOWLING, C.E. **1/0**
* Or with "The Elements of Plane Trigonometry," in One Volume, 2s. 6d.

Differential Calculus,
Elements of the. By W. S. B. WOOLHOUSE, F.R.A.S., &c. . . **1/6**

Integral Calculus.
By HOMERSHAM COX, B.A. . . **1/0**

Algebra,
The Elements of. By JAMES HADDON, M.A. With Appendix, containing Miscellaneous Investigations, and a Collection of Problems . . **2/0**

A Key and Companion to the Above.
An extensive Repository of Solved Examples and Problems in Algebra. By J. R. YOUNG **1/6**

Commercial Book-keeping.
With Commercial Phrases and Forms in English, French, Italian, and German. By JAMES HADDON, M.A. **1/6**

Arithmetic,
A Rudimentary Treatise on. With full Explanations of its Theoretical Principles, and numerous Examples for Practice. For the Use of Schools and for Self-Instruction. By J. R. YOUNG, late Professor of Mathematics in Belfast College. Eleventh Edition **1/6**

A Key to the Above.
By J. R. YOUNG **1/6**

Equational Arithmetic,
Applied to Questions of Interest, Annuities, Life Assurance, and General Commerce; with various Tables by which all Calculations may be greatly facilitated. By W. HIPSLEY **1/6**

Arithmetic,
Rudimentary, for the Use of Schools and Self-Instruction. By JAMES HADDON, M.A. Revised by ABRAHAM ARMAN **1/6**

A Key to the Above.
By A. ARMAN **1/6**

WEALE'S SCIENTIFIC AED TECHNICAL SERIES. 13

Mathematical Instruments:
Their Construction, Adjustment, Testing, and Use concisely Explained By J. F. HEATHER, M.A., of the Royal Military Academy, Woolwich. Fourteenth Edition, Revised, with Additions, by A. T. WALMISLEY, M.I.C.E. Original Edition, in 1 vol., Illustrated **2/0**

⁎ *In ordering the above, be careful to say "Original Edition," or give the number in the Series (32), to distinguish it from the Enlarged Edition in 3 vols. (as follows)—*

Drawing and Measuring Instruments.
Including—I. Instruments employed in Geometrical and Mechanical Drawing, and in the Construction, Copying, and Measurement of Maps and Plans. II. Instruments used for the purposes of Accurate Measurement, and for Arithmetical Computations. By J. F. HEATHER, M.A. . **1/6**

Optical Instruments.
Including (more especially) Telescopes, Microscopes, and Apparatus for producing copies of Maps and Plans by Photography. By J. F. HEATHER, M.A. Illustrated **1/6**

Surveying and Astronomical Instruments.
Including—I. Instruments used for Determining the Geometrical Features of a portion of Ground. II. Instruments employed in Astronomical Observations. By J. F. HEATHER, M.A. Illustrated. . . . **1/6**

⁎ *The above three volumes form an enlargement of the Author's original work, "Mathematical Instruments," price 2s. (Described at top of page.)*

Mathematical Instruments:
Their Construction, Adjustment, Testing and Use. Comprising Drawing, Measuring, Optical, Surveying, and Astronomical Instruments. By J. F. HEATHER, M.A. Enlarged Edition, for the most part entirely re-written. The Three Parts as above, in One thick Volume. **4/6**

The Slide Rule, and How to Use It.
Containing full, easy, and simple Instructions to perform all Business Calculations with unexampled rapidity and accuracy. By CHARLES HOARE, C.E. With a Slide Rule, in tuck of cover. Fifth Edition. . . **2/6**

Logarithms.
With Mathematical Tables for Trigonometrical, Astronomical, and Nautical Calculations. By HENRY LAW, C.E. Revised Edition . . . **3/0**

Compound Interest and Annuities (Theory of).
With Tables of Logarithms for the more Difficult Computations of Interest, Discount, Annuities, &c., in all their Applications and Uses for Mercantile and State Purposes. By FEDOR THOMAN, Paris. Fourth Edition . **4/0**

Mathematical Tables,
For Trigonometrical, Astronomical, and Nautical Calculations; to which is prefixed a Treatise on Logarithms. By H. LAW, C.E. Together with a Series of Tables for Navigation and Nautical Astronomy. By Professor J. R. YOUNG. New Edition **4/0**

Mathematics,
As applied to the Constructive Arts. By FRANCIS CAMPIN, C.E., &c. Second Edition **3/0**

Astronomy.
By the late Rev. ROBERT MAIN, F.R.S. Third Edition, revised and corrected to the Present Time. By W. T. LYNN, F.R.A.S. . . . **2/0**

Statics and Dynamics.
The Principles and Practice of. Embracing also a clear development of Hydrostatics, Hydrodynamics, and Central Forces. By T. BAKER, C.E. Fourth Edition **1/6**

14 WEALE'S SCIENTIFIC AND TECHNICAL SERIES.

BOOKS OF REFERENCE AND MISCELLANEOUS VOLUMES.

A Dictionary of Painters, and Handbook for Picture Amateurs.
Being a Guide for Visitors to Public and Private Picture Galleries, and for Art-Students, including Glossary of Terms, Sketch of Principal Schools of Painting, &c. By Philippe Daryl, B.A. **2/6**

Painting Popularly Explained.
By T. J. Gullick, Painter, and John Timbs, F.S.A. Including Fresco, Oil, Mosaic, Water Colour, Water-Glass, Tempera, Encaustic, Miniature, Painting on Ivory, Vellum, Pottery, Enamel, Glass, &c. Fifth Edition **5/0**

A Dictionary of Terms used in Architecture, Building, Engineering, Mining, Metallurgy, Archæology, the Fine Arts, &c.
By John Weale. Sixth Edition. Edited by Robt. Hunt, F.R.S. Numerous Illustrations **5/0**

Music:
A Rudimentary and Practical Treatise. With numerous Examples. By Charles Child Spencer **2/6**

Pianoforte,
The Art of Playing the. With numerous Exercises and Lessons. By Charles Child Spencer **1/6**

The House Manager.
Being a Guide to Housekeeping, Practical Cookery, Pickling and Preserving, Household Work, Dairy Management, Cellarage of Wines, Home-brewing and Wine-making, Stable Economy, Gardening Operations, &c. By An Old Housekeeper **3/6**

Manual of Domestic Medicine.
By R. Gooding, M.D. Intended as a Family Guide in all cases of Accident and Emergency. Third Edition, carefully revised . . **2/0**

Management of Health.
A Manual of Home and Personal Hygiene. By Rev. James Baird **1/0**

Natural Philosophy,
For the Use of Beginners. By Charles Tomlinson, F.R.S. . . **1/6**

The Electric Telegraph,
Its History and Progress. With Descriptions of some of the Apparatus. By R. Sabine, C.E., F.S.A., &c. **3/0**

Handbook of Field Fortification.
By Major W. W. Knollys, F.R.G.S. With 163 Woodcuts . . **3/0**

Logic,
Pure and Applied. By S. H Emmens. Third Edition . . **1/6**

Locke on the Human Understanding,
Selections from. With Notes by S. H. Emmens . . **1/6**

The Compendious Calculator
(*Intuitive Calculations*). Or Easy and Concise Methods of Performing the various Arithmetical Operations required in Commercial and Business Transactions; together with Useful Tables, &c. By Daniel O'Gorman. Twenty-seventh Edition, carefully revised by C. Norris . . **2/6**

WEALE'S SCIENTIFIC AND TECHNICAL SERIES. 15

Measures, Weights, and Moneys of all Nations.
With an Analysis of the Christian, Hebrew, and Mahometan Calendars.
By W. S. B. WOOLHOUSE, F.R.A.S., F.S.S. Seventh Edition . **2/6**

Grammar of the English Tongue,
Spoken and Written. With an Introduction to the Study of Comparative Philology. By HYDE CLARKE, D.C.L. Fifth Edition. . . . **1/6**

Dictionary of the English Language.
As Spoken and Written. Containing above 100,000 Words. By HYDE CLARKE, D.C.L. **3/6**
Complete with the GRAMMAR **5/6**

Composition and Punctuation,
Familiarly Explained for those who have neglected the Study of Grammar.
By JUSTIN BRENAN. 18th Edition. **1/6**

French Grammar.
With Complete and Concise Rules on the Genders of French Nouns. By
G. L. STRAUSS, Ph.D. **1/6**

English-French Dictionary.
Comprising a large number of Terms used in Engineering, Mining, &c.
By ALFRED ELWES **2/0**

French Dictionary.
In two Parts—I. French-English. II. English-French, complete in One Vol. **3/0**
*** Or with the GRAMMAR, **4/6**.

French and English Phrase Book.
Containing Introductory Lessons, with Translations, Vocabularies of Words,
Collection of Phrases, and Easy Familiar Dialogues **1 6**

German Grammar.
Adapted for English Students, from Heyse's Theoretical and Practical Grammar, by Dr. G. L. STRAUSS **1/6**

German Triglot Dictionary.
By N. E. S. A. HAMILTON. Part I. German-French-English. Part II.
English-German-French. Part III. French-German-English . . **3/0**

German Triglot Dictionary.
(As above). Together with German Grammar, in One Volume . **5/0**

Italian Grammar.
Arranged in Twenty Lessons, with Exercises. By ALFRED ELWES. **1/6**

Italian Triglot Dictionary,
Wherein the Genders of all the Italian and French Nouns are carefully noted down. By ALFRED ELWES. Vol. I. Italian-English-French. **2/6**

Italian Triglot Dictionary.
By ALFRED ELVES. Vol. 2. English-French-Italian. . **2/6**

Italian Triglot Dictionary.
By ALFRED ELWES. Vol. 3. French-Italian-English. . . **2/6**

Italian Triglot Dictionary.
(As above). In One Vol. . . . **7/6**

Spanish Grammar.
In a Simple and Practical Form. With Exercises. By ALFRED ELWES **1/6**

Spanish-English and English-Spanish Dictionary.
Including a large number of Technical Terms used in Mining, Engineering, &c., with the proper Accents and the Gender of every Noun. By ALFRED ELWES **4/0**
*** Or with the GRAMMAR, **6/0**.

16   WEALE'S SCIENTIFIC AND TECHNICAL SERIES.

Portuguese Grammar,
In a Simple and Practical Form. With Exercises. By ALFRED ELWES   **1/6**

Portuguese-English and English-Portuguese Dictionary.
Including a large number of Technical Terms used in Mining, Engineering, &c., with the proper Accents and the Gender of every Noun. By ALFRED ELWES. Third Edition, revised **5/0**
*_** *Or with the* GRAMMAR, **7/0**.

Animal Physics,
Handbook of. By DIONYSIUS LARDNER, D.C.L. With 520 Illustrations. In One Vol. (732 pages), cloth boards **7/6**
*_** *Sold also in Two Parts, as follows:—*
ANIMAL PHYSICS. By Dr. LARDNER. Part I., Chapters I.—VII.   **4/0**
ANIMAL PHYSICS. By Dr. LARDNER. Part II., Chapters VIII.—XVIII.   **3/0**

www.ingramcontent.com/pod-product-compliance
Lightning Source LLC
Chambersburg PA
CBHW021830230426
43669CB00008B/921